# The Proper Pugilist

# The Proper Pugilist

*Essays on the Milling Art*

Roger Zotti

Copyright © 2015 by Roger Zotti.

ISBN:      Softcover         978-1-5144-1707-2
           eBook             978-1-5144-1706-5

All rights reserved. No part of this book may be reproduced or transmitted in any form or by any means, electronic or mechanical, including photocopying, recording, or by any information storage and retrieval system, without permission in writing from the copyright owner.

Any people depicted in stock imagery provided by Thinkstock are models, and such images are being used for illustrative purposes only.
Certain stock imagery © Thinkstock.

Print information available on the last page.


Rev. date: 12/23/2015



**To order additional copies of this book, contact:**
Xlibris
1-888-795-4274
www.Xlibris.com
Orders@Xlibris.com
726990

# Contents

Author's Note ..................................................................... xi
Acknowledgements ............................................................ xiii
Preface ............................................................................. xv
Introduction .................................................................... xvii

Rocky and the Joe Louis Look-alike ........................................ 1
The Legend of Billy Miske ..................................................... 5
My First Professional Fight ................................................... 7
Thanks, Uncle Cheech ........................................................ 10
Fistic and Film Comebacks ................................................. 14
Thirteen Seconds .............................................................. 17
*Ring* (August 1951) .......................................................... 20
Diamond's Fearless Heart .................................................. 23
Uncle Cheech Talks Abeetz ................................................ 26
Cut and Shoot ................................................................... 30
The Milling Art ................................................................. 32
Toledo Slaughter .............................................................. 35
Gentleman Jack ................................................................ 38
Dempsey According to Roger Kahn .................................... 41
Don't Crowd the Champ .................................................... 44
Stanley the Man ................................................................ 47
The Proper Pugilist ........................................................... 51
Revisiting Kaletsky, Silver, and Vitale ................................. 57
Where Have You Gone, Court Sheppard? ........................... 62
Sonny's Visit .................................................................... 65
Did Harry Save Rocky? ...................................................... 68
Joe and Terrible Tony ....................................................... 71
Long Overdue .................................................................. 75
Sources ........................................................................... 79

To Maryann, Tom, Leslie, Katja, Roy, Tee, Ryan, Tillie, and Jake.

Boxing is the most basic and fundamental of all sports and its records will never be without their nobler aspects.
Gene Tunney

I don't train. The way I look at it is if you train, you're going to be in condition to go more rounds. If you go more rounds you're going to have to sustain more punishment. So it's not worth it for me to train . . . I don't have the ability to inflict pain on contenders or potential contenders without the use of foul tactics.
Bruce "The Mouse" Strauss

A young man, Rocky Marciano, knocked the old man [Joe Louis] out . . . An old man's dream ended. A young man's vision of the future opened wide. Young men have visions, old men have dreams. But the place for old men to dream is beside the fire.
Red Smith

I'm a prizefighter. I'm like Willie Pep. Willie Pep never liked taking punches.
Charlotte Rampling (*Broadchurch*/ second season)

Joe Louis was the most beautiful fighting machine I have ever seen.
Ernest Hemingway

# Author's Note

In *Portrait inside My Head*, Phillip Lopate's latest collection of critical and personal essays, he says that the "shagginess" of the essay form—"its discontinuous forms of consciousness"—fascinates him. Well, it fascinates me, too, which is why the essays in *The Proper Pugilist* are shaggy.

I wrote *The Proper Pugilist* because I wanted to share my thoughts on what George Plimpton, in *Shadow Box: An Amateur in the Ring*, calls that "strange but interesting fraternity—that of the world of boxing." Also, I wanted to express my ideas about the sport and its courageous participants.

Sit back and read, remember and enjoy!

# Acknowledgements

Many thanks to my wife Maryann and Rick Kaletsky for their editorial assistance; to Phil Carney for suggesting the book's title and writing the introduction; and to Dan Cuoco, editor, director, and publisher of the *International Boxing Research Organization Journal (IBRO)* for his assistance.

# Preface

On September 22, 1927, at Soldier Field, Chicago, former heavyweight champion Jack Dempsey, behind on the officials' scorecards, knocks down world heavyweight champion Gene Tunney in the seventh round.

In Tunney's article, "The Long Count," he admits that he "never glimpsed that savage left hook which Jack swung on my jaw. Getting hit is commonplace, but not seeing it coming really injured my pride. I was always cocksure about my eyesight in the ring . . . The blow was the second in that series of seven that put me on the canvas for the first time in my life."

Dempsey rests against the ropes not far from the floored Tunney. Seconds pass. During this time referee Dave Barry shouts at Dempsey: "Go to a neutral corner, Jack! A neutral corner!"

Finally Dempsey obeys.

Barry doesn't pick up timekeeper Paul Beeler's count, which has reached four. Instead, he begins his own. Because of Dempsey's stubbornness, hot-headedness, or confusion (maybe all three), Tunney has about four extra seconds to recover.

"Could I, in that space of time, have got up and carried on as I did?" Tunney writes in "My Fights with Jack Dempsey." "I'm quite sure I could have. When I regained consciousness after the brief period of black-out, I felt I could have jumped up immediately and matched my legs against Jack's just as I did."

So, back on his feet, Tunney retreats. Dempsey can't catch him. His legs are "so heavy that he couldn't move with any agility at all, and I was able to hit him virtually at will. He was almost helpless when the final bell rang—sticking it out with stubborn courage."

# Introduction

Poised to strike,
Thus does engage—
The proper pugilist
—amaze
With skills, wrought of desire,
The body moves—
        The soul afire
And like the mongoose
        With the cobra
The end result
        Proves truly sober

Philip W. Carney

# Rocky and the Joe Louis Look-alike

*Change*

In *The Real Rockys: A History of the Golden Age of Italian Americans in Boxing 1900-1950,* Rolando Vitale writes about Rocky Marciano's name change. Al Weill, Rocky's manager, recommended the change because the fighter's surname, Marchegiano, "had been carved up repeatedly in the local and national press."

According to Vitale, Weill said let's try Rocky Mack. It's easy to remember. Easy to spell. Easy to pronounce. But it didn't sound Italian enough for Rocky. Then Weill proposed Rocky March. Easy and simple. That wasn't Italian enough for Rocky, either.

Listening to the conversation was a Rhode Island promoter, who suggested removing the letters H and E and G: What's left, he said, is M-A-R-C-I-A-N-O.

Rocky nodded. Though it sounded Italian, Rocky's biggest worry was his father: Would he approve of the name change?

*Defeats and Victory*

As an amateur Rocky was defeated four times. In 1946, fighting under his real name, he lost to Henry Lester, an experienced amateur heavyweight. In the opening session Lester jabbed Rocky, overweight and sluggish, silly. In round two Rocky, angry and frustrated, kneed Lester in the balls and was disqualified.

In his dressing room after the fight, he told his brother Peter that he'd never again be out of shape for a fight. He never was.

Another defeat came on March 2, 1948, in a Golden Gloves bout at the Ridgewood Grove Arena in Brooklyn, New York, where Rocky, now fighting as Rocky Marciano, fought highly touted heavyweight Coley

Wallace. After Rocky battered Wallace for three rounds, his handlers and fans were certain he earned the decision. Not so! The verdict went to Wallace. Rocky's supporters were irate. Some even threatened the officials.

Writing for *The Lowell News*, sports editor John F. Kearney called the decision "one of the most putrid . . . handed down from a Golden Gloves tournament ring anywhere—and this writer has been watching boxing bouts since Jack Benny owned a Maxwell . . ."

As a professional Marciano went undefeated in forty-nine fights. Only Don Mogard, Ted Lowry (twice), Willis Applegate, Roland LaStarza, and Ezzard Charles went the distance with him. (In other words, 87.5 per cent of his opponents went nighty-night before their accustomed bedtime.)

W.C. Heinz begins "Brockton's Boy," his piece about the impact of Marciano's victory over Jersey Joe Walcott on the shoe manufacturing city of Brockton, Massachusetts, like this: "On September 23, 1927 . . . Mr. Fred Denly. . . succeeded in bringing to bloom a two-headed dahlia. Twenty-five years later, to the very day, Mr. Rocco Marchegiano, of 168 Dover Street, same city, distinguished himself in another field. In Philadelphia he hit Mr. Arnold Cream, of 1020, Cooper Street, Camden, New Jersey, on the chin and won the heavyweight championship of the world."

Two examples—among many others—illustrate how ablaze the Brockton folks were on the night of the title fight. First, the Brockton Police Department's night shift picked up a blow-by-blow description of it, on shortwave radio, from a Canadian station. But the description was in French. And none of the cops spoke or understood French.

"They moved the radio back to the cell where a Bridgewater prisoner of French extraction had provided them with a translation," Heinz writes. "The next court, the bilingual benefactor was fined five dollars for drunkenness."

Second, consider young Nicholas Rando. He lived on the street where Rocky, as a youngster, delivered the Brockton Enterprise-Times. Nicholas "will forever be lacking the first joint of two fingers of his right hand. When Marciano, with his own right hand, flattened Walcott, the excitement that raged through Brockton's Ward Two Memorial Club reached such a pitch that Rando, then fourteen years old, fed his hand into a ventilating fan."

*Heinz concludes: "It is safe to say that as long as there is boxing and a Brockton twenty miles due south of Boston, the impact of the punch with which Rocky Marciano . . . knocked out Jersey Joe Walcott . . . will be felt."*

### Coley and Joe

Handsome Coley Wallace turned professional in 1950 and retired in 1956. At one point in the early fifties *Ring* magazine, boxing's bible, ranked him seventh in the heavyweight division.

I recall his nationally televised fight against cagey Jimmy Bivins in New York's St. Nick's Arena, on September 19, 1952. Wallace was ahead for eight rounds. But in the ninth Bivins, outweighed by fifteen pounds, knocked him cold with a perfectly timed right hand to the jaw.

According to boxing historian Jerry Fitch, in *James Louis Bivins: The Man Who Would Be Champion*, "Wallace had a habit of hitching up his trunks when the bell rang . . . and Jimmy took advantage of that bad habit by charging across the ring when the bell sounded and launching a right hand that put Coley down and out . . ."

Because he resembled former heavyweight titleholder Joe Louis, Wallace was cast in a film about the life of the great Brown Bomber. Titled *The Joe Louis Story*, the low budget 1953 film co-starred veteran actor Paul Stewart, who praised Wallace's acting.

I saw the movie at the Roger Sherman Theater in New Haven. Its biggest fault was its background score. Too intrusive. Sometimes you struggled to hear the dialogue.

Wallace did a good job as Louis. Some athletes appear in a movie and memorize their lines, then recite them with no feeling. Not Coley. He spoke his lines naturally and, when needed, with emotion. He had an infectious smile, too.

A key scene takes place between Louis and Hilda Simms, a good actress. She plays Louis' wife, Marva Trotter Louis, and wants him to break training for his second fight with Max Schmeling. It's his daughter's birthday and she's having a party. He tells her he can't attend because he's in training. Throughout the movie Louis is depicted as a nice guy, which he was. There was nothing arrogant about him. But his profession came first.

James Edwards played Jack "Chappie" Blackburn, Joe's trainer. Blackburn was a fierce individual. Supposedly he once murdered a man. A terrific actor, Edwards captured Blackburn's ferocity. Had Edwards been white he would've been a big star.

**Joe DeAngelis and Bob Girard were the other amateur fighters who defeated Marciano.**

# The Legend of Billy Miske

1

On January 1, 1924, "Billy [Miske died at] 8:00 PM . . . with [his wife] Marie and his manager and friend, Jack Reddy, by his bedside," Clay Moyle writes in *Billy Miske: The St. Paul Thunderbolt*. "His long fight was finally over."

Shortly before he died Miske told his wife to thank Jack Dempsey for giving him a shot at the title.

Moyle's *Billy Miske,* an empathetic, well-crafted biography, is about an outstanding, courageous prizefighter diagnosed with Bright's disease in July 1919. With a wife and two children to support and his medical bills mounting, Miske was in financial trouble. Realizing his boxing career might be over, he took a chance and invested in an automobile business. The venture put him deeper in debt.

But with the disease in remission, Miske did what he did best to earn money: He kept fighting.

Enter heavyweight champion Jack Dempsey. Miske wrote a letter to him and said that a title bout would help him financially. Dempsey and his manager, Jack Kearns, agreed to the fight. (Miske's share of the gate was about twenty-five thousand dollars, his biggest payday ever.)

Before Dempsey became champion, he fought Miske twice in 1918. The first fight was a draw, while Dempsey won the second by decision. Dempsey had been knocking out almost every opponent he faced, but he couldn't stop Miske. Couldn't even floor him.

Their third fight was held on Labor Day, September 6, 1920, in Benton Harbor, Michigan, at a minor league baseball stadium. Miske's downfall began early. In the first round, Dempsey landed a fierce right hand punch under the challenger's heart. Miske went down in pain.

5

"It was the first time in eight years of fighting, and more than 80 professional bouts," Moyle writes, "that Billy had ever been off his feet." Struggling to his feet, he survived the round.

After Miske was floored in the third round, Dempsey positioned himself in a corner behind him. When the St. Paul fighter rose to his feet, the champion "moved a fraction to the side and delivered a brutal right hook to the jaw that sent Billy down and out, one minute and thirteen seconds into the round . . . The scene would repeat itself three years later in a wild heavyweight championship affair between Dempsey and Argentina's Luis Firpo before 80,000 fans at New York's Polo Grounds." Firpo was dropped seven times in the opening session and Dempsey "was permitted to stand over the fallen fighter and immediately knock him down again the moment he rose to his feet."

In his book, *Tunney: Boxing's Brainiest Champ and His Upset of the Great Jack Dempsey*, Jack Cavanaugh quotes Dempsey, who phoned Firpo a few days after the fight and apologized for belting him when he was barely back on his feet: "You hit me so hard I didn't know what I was doing. I was mixed up."

Firpo's response was that there were three men in the ring, and if Dempsey didn't know what he was doing, why didn't he hit the referee?

### 3

After his loss to Dempsey, Miske continued fighting, winning twenty-one of twenty-three bouts. His last fight was against old foe Bill Brennan in Omaha, Nebraska, at the City Auditorium, on November 7, 1923. Though the disease had recurred and Billy was dying, he told reporters he was in top shape for the fight. "The truth was Billy's body wouldn't have been able to handle the strain of the type of training he would normally have undergone before a bout," Moyle says.

Miske knocked out Brennan in the fourth round. It was only the second time in his one hundred and two bout career he had been stopped.

Miske used guts and determination and character, along with his boxing skills, to defeat Brennan.

***The Legend of Billy Miske: The Saint Paul Thunderbolt*** **was first reviewed in the *International Boxing Research Organization Journal* (issue 123/ September 2014).**

# My First Professional Fight

When Natale Menchetti became a prizefighter in 1934, he changed his name to Nathan Mann. Born in New Haven, Connecticut, he fought professionally for fourteen years. In 1999 he passed away at age eighty-four. Seven years later he was inducted into the Connecticut Boxing Hall of Fame.

In 1938 world heavyweight champion Joe Louis was in his prime when he defended his title against Mann at Madison Square Garden. Mann stunned the great Louis with a right cross to the jaw in the opening round. In the third round, Louis' punching power was too much for Mann and the game challenger was knocked out.

One year before he retired, Mann fought Bill Weinberg. I was nine years old and saw the fight, at the New Haven Arena, with my grandfather and my uncle Vincenzo, diehard Mann fans. It was the first of many fights I'd see in person.

Of course I was thrilled, but food was my number one priority. "Does the Arena have hotdogs?" I asked my uncle. "Candy apples? Cotton candy? Popcorn?" (All the major food groups.)

When Weinberg entered the ring, the booing was long and loud. No surprise. When Mann, a hero to New Haven's large Italian population, walked down the aisle he was welcomed with a thunderous ovation, prompting my grandfather to remove the evil-smelling cigar from his mouth and utter, "Weinberg's too slow. Natie'll flatten him real quick. You just watch, kid." Then he jammed his cigar confidently back into his mouth.

My uncle Vincenzo, the world's most energetic person, nodded in agreement, lit a Raleigh and inhaled. (Two days later he exhaled.)

Good old grandpa and Uncle Vincenzo were wrong. Weinberg closed Mann's eye in the fourth round and proceeded to outpunch the local favorite. When the tenth round ended, it was clear who had won. But boxing is notorious for hometown decisions, and the referee, the sole official, awarded the fight to Mann. Most of the crowd cheered, including yours truly. But deep down we knew and Mann knew and his crew knew that Weinberg deserved the nod.

They fought again three months later at the Arena. This time Mann, better conditioned than in their first bout, dropped Weinberg in the second round and boxed his way to a unanimous decision.

On June 28, 1948, Mann fought Bridgeport's Bernie Reynolds, a young hard hitting heavyweight prospect, at the Arena. They battled for the USA New England Heavyweight title. (Don't ask.) Mann was stopped in four rounds. That was his last fight.

*One of Bernie Reynolds' best fights was against Rocky Marciano, in 1952, at the Rhode Island Auditorium. In the first round Reynolds outslugged and out-boxed Rocky. Then he became careless.*

*Near the end of the second round, Rocky started finding the range. In the third round he hit Reynolds with a right hand shot similar to the one he knocked out Jersey Joe Walcott with years later. Curtains for Reynolds.*

*Maybe if Reynolds had been handled better, he might've gone further in his career. He packed a wallop and had good boxing skills.*

*Some people remember Reynolds for his fist fight with actor Robert Mitchum. In 1952, Mitchum was making an RKO movie titled* One Minute to Zero, *and one afternoon he, co-star Charles McGraw, and a few pals from the film crew went into town from their location in Colorado Springs. They were ready for some movie star drinking at the Red Fox Lounge. An Army private walked in and McGraw, brave with booze, certainly not uncommon for most lushes, told him he was dressed like a slob. They started pushing and shoving. Of course McGraw didn't know the private was a professional boxer.*

*Mitchum became involved in the melee. When it was over Reynolds had been injured. A concussion. When Mitchum's press agent learned about the fight, he got the word out that the actor and a heavyweight professional boxer named Bernie Reynolds fought in a bar, and Mitchum knocked him out. It made great copy and gave the actor's macho image a boost at the box office.*

*Maybe that's what happened. Maybe not. Maybe Mitchum's pals, including McGraw, jumped Reynolds. Reynolds' side of the story was never told.*

A decent heavyweight, Mann gave the Italian people of New Haven someone to cheer for—just like Julie Kogon, a very good lightweight who fought from 1937 to 1950, gave the city's Jewish folks someone to root for. There was plenty of anti-Italian and anti-Jewish sentiment in the city—in the country, too, which kids didn't learn about in their history books back then.

As for Weinberg, he was a fearless journeyman and always provided a good test for upcoming fighters. He came from Chelsea, Massachusetts, and fought Bernie Reynolds in 1948, the same year Mann did, at Municipal Stadium in Waterbury. Reynolds stopped him in the eighth round.

His opposition was better than Mann's. He was in against guys like Art Henri, Bob Baker, Lee Oma, Elkins Brothers, Cesar Brion, Roland LaStarza, and Embrel Davidson. The fight against Davidson was his last. He was stopped in the tenth.

Unlike Mann, Weinberg never fought one of the greatest heavyweights ever: Joe Louis.

# Thanks, Uncle Cheech

"If you ever write a book," Uncle Cheech said, "include a chapter devoted to a non-boxing event and a boxing event that happened during the Roaring Twenties. Must be concisely written. No padding, kiddo."

(That's Uncle Cheech speaking. Everyone should have such an uncle. A book should be written about him, though he wasn't famous. Just consistent and self-educated.)

"Yes," I reply. "Yes, indeed, and I'll give you full credit for the idea. Thanks, Uncle Cheech."

I saw Uncle Cheech three days later. "The book is a history classic," I tell him. "It's Frederick Lewis Allen's *Only Yesterday: An Informal History of the 1920's*; and the non-boxing event is—the Floyd Collins tragedy. The boxing event—what else?—is Tunney vs. Dempsey, their rematch."

"Read it a long time ago," he replies. "Good choice."

Collins was a young Kentuckian caught in a cave-in, and a reporter, W.B. Miller of the *Louisville Courier-Journal*, crawled into the cave and interviewed him. Miller's series of dispatches held the country fascinated because, according to Lewis, "Collins' plight contained those elements of dramatic suspense and individual conflict with fate which make a great news story, and every city editor, day after day, planted it on page one."

Then on February 17, 1925, Collins, trapped for eighteen days, died. About a month later there was another cave-in, this time in a North Carolina mine, where "71 men were caught and 53 actually lost their lives. It attracted no great notice. It was 'just another mine disaster.' Yet for more than two weeks the plight of [Collins] riveted the attention of the nation on Sandy Cave, Kentucky."

*Nearing the peak of his career, Kirk Douglas starred in 1951's* Ace in the Hole, *a film loosely based on the Floyd Collins tragedy. The great Billy Wilder directed. Along with Lesser Samuels and Walter Newman, he wrote the screenplay.*

*In* The Ragman's Son, *Douglas' autobiography, Douglas writes that before filming began he approached Wilder and said, "Billy, great script. I love it. My character—maybe I should make him a little softer. Make him at least a little sympathetic."*

*Wilder disagreed. "Kirk, listen to me. Go for it. Right from the start I vant you to make the guy talk, act, and look like the unscrupulous bastard he iz. Make him someone dat the audience vill hate."*

*And Douglas did! It doesn't take long before viewers realize what an ambitious heel* Douglas' Chuck *Tatum, a reporter, is: He deliberately keeps the trapped man down in the mine so he can continue filing stories.*

*The real W.B. Miller won a Pulitzer Prize for his reporting and wasn't anything like Douglas' character. "Only a few people might have heard of Collins's predicament if W.B. Miller," Allen writes, "had not been slight of stature, daring, and an able reporter . . ." Miller volunteered to enter the cave, delivering food and water to Collins. There was no attempt to delay the rescue.*

*When the film was first released, Douglas says, it received unfavorable reviews because it's about people who enjoy stopping and "staring at accidents."*

*In* Kirk Douglas: The Man and the Actor, *Michael Munn notes that "by 1951 standards, the film was too harrowing a movie to be entertaining, and the public could not be enticed." At the same time, the press panned it because of "the way they were portrayed in the picture."*

*But with the passage of time—and a title change to* The Big Carnival*— it has, Douglas writes, "become an underground classic, playing at revival theaters."*

The boxing event took place about two years after the Collins and North Carolina tragedies: It was the return fight between Jack Dempsey and heavyweight champion Gene Tunney at Soldier Field, Chicago, in 1927, before 145,000 fans. (In 1926, Tunney had dethroned Dempsey before 130,000 fans at Sesquicentennial Stadium, in Philadelphia.)

Again Tunney won by unanimous decision. Its indelible moment took place in round seven when, Lewis writes, "Tunney fell and the referee, by delaying the beginning of the count until Dempsey had reached his corner, gave Tunney some thirteen seconds to recover . . ."

Speaking with reporters after the fight, Tunney said that, yes, the referee started the count after Dempsey was in a neutral corner. But that was the right thing to do. He was following the rules. There shouldn't be any controversy.

*In* All Those Mornings . . . at the Post, *Shirley Povich writes of the long count that "Gene Tunney could have risen from the canvas at the count of five had he so elected." He had "[regained] his composure." And back on his feet, "Gene Tunney played the game like a champion, taking full advantage of the rules that protected him, waiting until the count of nine had been reached before he went back to the battle."*

*In rounds eight, nine, and ten he even outslugged Dempsey, as if he wanted to knock him out.*

*Consider Povich's 1950 column for the* Post, *"Dempsey or Louis: Who Was Better?" Povich speculates that had Dempsey and Louis fought in the same era, "instead of attaining their peaks twenty years apart," Dempsey would've emerged victorious, probably by knockout.*

*Louis' Achilles heel was his chin: Jimmy Braddock, Max Schmeling, Buddy Baer, Tony Galento, Joe Walcott, and Rocky Marciano had him down. Tami Mauriello knocked him across the ring with a right hand in the first round of their 1946 title bout at Yankee Stadium, "and it appeared," Povich writes, "as if a new champion was in the making." Always in superb condition and never one to panic, Louis regrouped and "knocked Mauriello out in the same round with a punch that ranked as one of the most explosive of all time . . ."*

*In his* Post *column, in 1969, Povich writes that Dempsey in his prime "would lick anybody of that day or any other day. Louis, Marciano, Clay—you name them. Dempsey was better."*

*Still, he adds, Gene Tunney "was the most sharp-shooting puncher of them all. He punched to spots." And yes, he could take a shot.*

*And maybe he could've taken Joe Louis, but "the most reasonable belief is that Louis would have caught up with Tunney somewhere within 10 rounds . . . That's just an idea. Suppose you take it from there."*

*Because you asked, my pick is Louis over Dempsey. Probably by a technical knockout. He hit just as hard—probably harder—than Dempsey. He was a vastly superior boxer. Louis' strategy would be to stay away from the Manassa Mauler for four or five rounds—and he'd accomplish with his left jab, one of the most potent in boxing history.*

*Though Dempsey suffered only one knockout loss in his career, his defensive skills were minimal. By round nine or ten Louis' jab, coupled with the accuracy of his combination punching, would've closed and seriously cut Dempsy's eyes.*

# Fistic and Film Comebacks

*Swanson, Brando, and Rourke*

Light heavyweight champion Archie Moore came back from four knockdowns, three in the first round, to stop Canadian challenger Yvon Durelle in the eleventh. That happened in 1958. Trailing heavyweight champion Jersey Joe Walcott in their title bout, Rocky Marciano knocked him out in the 13$^{th}$ round with a devastating right hand to the jaw. That happened in 1952.

Comparable to the comebacks of Moore and Marciano are those of film actors Gloria Swanson, Marlon Brando, Mickey Rourke, and Frank Sinatra.

A huge silent movie star during the nineteen-twenties, Swanson was absent from films from 1934's *Music in the Air* to 1941's *Father Takes a Wife*. Then nine years later, age fifty, she returned to the screen in Billy Wilder's *Sunset Boulevard* (1950), portraying washed-up silent screen actress Norma Desmond. Her unforgettable performance earned her a Best Actress nomination, her third.

Jeanine Basinger writes in *Silent Stars* that "the movie was an enormous success for everyone involved, and for Swanson it meant she was back in the limelight again . . . Few actresses ever find a role like Norma Desmond. In truth, only one did, and she more than met the challenge . . . Yet the role of Norma Desmond was so powerful, and depicted such an obsessed character, that perhaps it doomed her. People began to think that Gloria Swanson was Norma Desmond. She never quite shook off Norma's shadow." *Airport 1975* (1974) was her final role and—yes—she played herself. "What could be more perfect than that?"

According to Patricia Bosworth in *Marlon Brando*, Paramount Pictures President Stanley Jaffe and chief of production Robert Evans "ridiculed the idea of Brando" portraying Don Vito Corleone in *The Godfather.* They demanded that he take a screen test.

Director and screenwriter Francis Ford Coppola, certain Brando wouldn't accede to the executives' wishes, tricked the actor: "He would tell the forty-seven-year old Brando he wanted to photograph him in makeup to test whether he could really pass for a sixty-five-year-old Italian man." Brando's screen test was a success. His performance earned him 1972's Best Actor. (It was his second Oscar. In 1954 he took home the eight pound statuette for his commanding turn as former boxer Terry Malloy in *On the Waterfront.*)

A huge star in the eighties, the talented and temperamental Rourke rejuvenated his career with *The Wrestler* (2008). "The film turned everything around for Mickey Rourke," writes Sandro Moretti in *Mickey Rourke: Wrestling with Demons,* "transforming him from a has-been to hot property . . ."

Initially, Warner Bros. was set to produce the film, but after learning that Rourke would be playing the lead, over-the-hill wrestler Randy "The Ram" Robinson, the studio "lost interest in the movie," claiming that "Mickey Rourke was no longer a box-office name who would guarantee a return on their investment." Wild Bunch, a French film company, stepped in and "would make their money back many times over."

Rourke didn't win the Oscar—which he deserved—but did take home a Golden Globe for 2008's Best Actor.

### *Sinatra*

Filmdom's biggest comeback belongs to Frank Sinatra for his unforgettable performance as Angelo Maggio in 1953's *From Here to Eternity.* In early 1950, his career as a singer and musical comedy star was at an all-time low. His wife Barbara had divorced him. Columbia Records and MGM Studios added to his woes by dropping him.

In April 1950, while performing at New York's Copacabana Night Club, he suffered a vocal cord hemorrhage.

Sinatra was down—but not out!

In 1952, Ava Gardner, his wife at the time, used her star power to persuade Columbia Pictures' dictatorial President Harry Cohn to give the singer a chance to test for the role of Maggio in the screen version of *From Here to Eternity,* James Jones' best-selling novel. She knew that

Sinatra had read the book several times and believed he was born to play Maggio.

The film's director, Fred Zinneman, and its producer, Buddy Adler, didn't want Sinatra; Eli Wallach was their first choice, but he was committed to another project.

Sinatra was required to do a screen test. It proved to be extraordinary. " . . . Frank improvised the saloon scene in which Maggio shakes dice and then casts them across a pool table," J. Randy Taraborrelli writes in *Sinatra: Behind the Legend*. "Frank used olives instead of dice . . . But that one gimmick wasn't all there was to Frank's performance. In fact, according to all accounts it was a commanding audition in every sense, part realistic, part theatrical . . ."

Opening on August 5, 1953—at the Capitol Theater in New York, Sinatra and the film— which also starred Burt Lancaster, Montgomery Clift, Deborah Kerr, and Donna Reed—won commercial and critical acclaim.

Fast forward to March 25, 1954, the Pantages Theater in Los Angeles. Nominated for 1953's Best Supporting Actor, Sinatra had some heady competition from Brandon de Wilde (*Shane*), Eddie Albert (*Roman Holiday*), Jack Palance (*Shane*), and Robert Strauss (*Stalag 17*). When Sinatra was announced as the winner, Taraborelli would write, "It was astonishing, in retrospect, that Frank Sinatra, basically known as a vocalist, would make one of the most dramatic comebacks in entertainment history as an actor."

# Thirteen Seconds

*Had It Won*

You had to feel for Laurent Dauthuille. If he had defeated Jake LaMotta in their 1950 title bout, he would've brought the middleweight championship back to France and been a national hero.

In 1949, Dauthuille won a unanimous ten round decision over LaMotta, at the Montreal Forum, in a non-title bout. For the rematch, a year later, eleven thousand fans saw Dauthuille, an 11-5 underdog, give champion LaMotta a hammering for fourteen rounds.

(Knowing he was far behind, LaMotta, in the round twelve, resorted to one of his oldest ploys, his possum act. After being hit by a right to the head, he began staggering across the ring, but Dauthuille didn't fall for it.)

Dauthuille, who had the fight won, came out slugging in the fifteenth.

Late in the round LaMotta spotted an opening and landed a left hook to the challenger's chin. Dauthuille staggered. Knowing he needed a knockout to win, LaMotta began throwing punches as hard and fast as he could. A whirlwind. Dauthuille went down. He tried to pull himself up by the middle strand, but the referee had reached ten. Thirteen seconds were left in the fight.

*Jake LaMotta became world middleweight champion when, on June 16, 1949, titleholder Marcel Cerdan was unable to come out for the tenth round. "The elevator muscle in the left shoulder had rendered his arm useless from the first round," Barney Nagler writes in "The Story of a Champion."*

*When and how did it happen? ". . . somewhere around the middle of the first round he wanted to go into a clinch but I didn't, and I was* half-punching, half-shoving him to get off me when he went down," LaMotta says in Raging Bull. *Was it a push? Did Cerdan slip? LaMotta didn't know.*

*The rematch was scheduled for September at the Polo Grounds, in New York, but LaMotta suffered a neck injury in training, and the fight was rescheduled for December at Madison Square Garden.* "On October 28," Nagler writes, "Cerdan emplaned for the United States. He never got there." *The plane crashed in the Azores, killing all forty-eight passengers on board.*

### *Excuses*

Excuse number one: In *Raging Bull* LaMotta writes that he was weakened because he had trouble making the middleweight limit for the fight. Excuse number two: LaMotta says he hurt his thumb in the fifth round and that hampered his attack. So instead of fighting his usual style—he usually fought in flurries—he boxed. Mistake.

What he had going for him in the last round was that Dauthuille "had been doing all the fighting for fourteen rounds, while I still had some energy left." So he charged the Frenchman, "throwing left and rights . . . and by pure luck he was wide open at the time . . . I followed him, throwing punches as hard as I could . . . I almost knocked him through the ropes, but he was a very game guy . . ."

### *No Excuses*

I had piles of boxing magazines stashed away at home. They were under my bed, under my pillow, under the dresser, and in my closet. I began researching LaMotta's fight against Dauthuille.

Nat Fleisher and his 1950 *Ring* magazine article to the rescue: "Dauthuille was one of the worst handled fighters I have ever seen in a championship fight. . . All Dauthuille had to do in the last round was stay away from LaMotta. The title was his. But his French handler kept shouting, 'Tear in. Fight him,' while Heinie Blattstein, Dauthuille's American second, begged the fighter to [just box and stay away from him]."

Also to the rescue was Raymond Alexander's 1950 *Boxing Illustrated* article, "The Bull's Moment of Truth": "Dauthuille had one thought—to avenge the defeat of his countryman Cerdan, who had since died in an airplane crash, and to take the crown back to France." At the same time, LaMotta, the first fighter to defeat Sugar Ray Robinson, wanted to retain his title. "But he also wanted a lot more—respect, adulation, and most important he wanted to regain the integrity he had lost in his fight with Billy Fox three long, painful years before."

In exchange for throwing the 1947 fight with Fox, LaMotta was promised a middleweight title shot by the lads who controlled the sport. That didn't happen until nine fights and two years later.

LaMotta offered no excuses about the Fox fight: "I don't know how long it's been since you stood up on a platform and listened to about twenty thousand people booing you, but I personally didn't like it. And the knowledge that I deserved it didn't help." He even saw humor in the situation, adding that "Dan Parker, the *Mirror* guy, said the next day my performance was so bad that he was surprised that Actors' equity didn't picket the joint."

# *Ring* (August 1951)

Back in the day I wasn't sensible enough to save my large collection of *Ring* magazines. If there was one I valued more than any other, it was the August 1951 edition. Why? Because my favorite fighter, New York middleweight Walter Cartier, graced its cover, and inside there was Barney Nagler's story about him.

Opening the magazine transported me back in time. Of course, I turned first to Nagler's piece, "He's in Love with Boxing." In it Cartier tells Nagler how important his upcoming 1951 fight at Madison Square Garden against ranking middleweight Gene "Silent" Hairston was. Important because local scribes never forgot Cartier's loss to middleweight Rocky Castellani in 1948. Reporters slammed him for folding up when things got too tough.

There was "a substantial basis for this criticism," Nagler writes. "Castellani had been battered by Cartier in the early rounds of their fight at St. Nick's, in New York, only to bounce back midway. He took over as boss and stopped Cartier."

So for Cartier the Hairston fight was a chance to redeem himself— and he did. Floored in the first and second rounds for counts of nine, he mounted a dramatic comeback. "He went out for the last round and Hairston greeted him with a two-fisted barrage," Nagler writes. "Cartier's senses were dulled. He fought out of instinct." The crowd roared. With a few seconds remaining he tagged "Hairston coming in with two rights to the jaw and a left." Hairston went down. The bell saved him from being counted out. Cartier was awarded a split decision. "He had pulled victory out of the fire."

*That August 1951* Ring, *priced at twenty-five cents, is in my possession again, thanks to Tom Doyle who gave it to me. I'm grateful and indebted.*

*I'd be remiss if I didn't mention its advertisements. Speaking of being transported back in time!* **"Here's the Kind of MEN I Build! Will You Let Me Prove I Can Make You a New Man?"** *(That's Charles Atlas speaking.) Takes only "15 Minutes a Day." Also included is a free forty-eight page illustrated book. "The book is a real prize for any fellow who wants a better build." Simply fill out the coupon and rush it to Mr. Atlas.*

*Don't overlook, of course, that nifty mouth protector from* **Everlast.** *Price is $2.50.* **"Greater Protection! Can't Fall out! Can't Gag! You Breathe Naturally! Fits Every Size!"**

*And if you like to pose almost naked in front of a mirror for your friends, check out "the new* **Coronado** *body mold* **'POSE-BREEFS.'** *They're worn "by leading body builders of the world!" You guessed it: they're reasonably priced at $5.95.*

Walter Cartier told Nagler that ever since he was a kid he wanted to be a fighter. But Vince, his identical twin brother, didn't and became a lawyer after they got out of the Navy. Walter never regretted becoming a professional prizefighter.

When Nagler writes that, "Walter and Vince react pretty much alike," I remembered W.C. Heinz's interview with Walter in *Once They Heard the Cheers*. In Heinz's piece, "The Same Person Twice," Walter, long retired from boxing, is quoted as saying when he and Vince went to a movie together, and "one of us wanted to go home, the other would have the same thought at the same time."

Heinz asks Vince if it was difficult watching his identical image being hit and learns that it was. He asks Walter if he and Vince ever bought each other the same Christmas or birthday present, and learns they never did because it would be too much like buying yourself a gift.

### *Golden Boy*

Synchronicity—that's what it was. A month before receiving the August *Ring*, I began thinking about those tough, journeyman fighters of the forties and fifties. Guys who never stopped coming forward. Guys like Artie Diamond, Harold Green, Charlie Salas, Billy Kilgore, Dick Wagner, Cary Mace, Jimmy Flood, and the most talented of the group, Jimmy Herring.

A promising middleweight of the early fifties, Herring, nicknamed "Golden Boy," fought too many good fighters too soon in his career. After I read Nagler's piece, I leafed through the magazine for other stories and discovered in Jersey Jones' column, "Broadcast from New York," a photo of the handsome Herring and a recap of his 1951 eight round decision over Artie Diamond of Middle Village, New York.

Seven months later he and Diamond clashed again, this time in a semi-final at Madison Square Garden. Herring came away with the decision. They fought twenty-one days later, again at the Garden, and Herring scored a fifth round TKO.

Herring hailed from Ozone Park, New York, and began fighting in 1948 and retired in 1954. He fought and lost to such stalwarts as Vic Cardell, Joey Giambra, Aldo Minelli, Ernie Durando, Billy Graham, Rocky Castellani, and Billy Kilgore. His most impressive win was a decision over Ralph "Tiger" Jones in 1952, but in the rematch a year later, Jones pounded out a unanimous decision.

# Diamond's Fearless Heart

*Old Friends*

In a Spanish night club in New York, one evening in 1976, Jose Torres, former light heavyweight champion turned writer, heard someone call his name. It was former prizefighter Artie Diamond. He and Torres embraced. "He looked happy and well," Torres writes. "He had not changed much with time." They talked. They drank. Torres learned Diamond was the club's "security officer"; that his first wife had died and he was remarried. They exchanged telephone numbers and promised to keep in touch.

Two weeks later, on February 23, trainer Cus D'Amato called Torres. He said that Artie Diamond was dead. A stranger had shot him in the heart. Died instantly. Diamond was forty-six years old.

Torres, author of "The Legend of Artie Diamond," was once trained by D'Amato, who later handled Floyd Patterson and Mike Tyson. Like Torres, they became world champions. D'Amato also trained Diamond.

*An Ear*

After Diamond's third loss to Jimmy Herring in 1951, Torres quotes D'Amato as saying he forced Artie to retire because he was a fighter "who got hit to prove he wasn't afraid. I couldn't, in good conscience, continue letting my fighter and friend keep taking punishment. His health and future were more important than boxing."

Later that year Diamond and several friends were involved in a botched robbery. Diamond shot and wounded a guard and was sentenced to seven to fifteen years in Green Haven, located in Stormville, New York. And true to the code of the streets, Torres writes, "the former fighter never divulged the names of his accomplices."

Diamond knew other inmates would test him. So he set out to prove that he wouldn't be pushed around. He wasn't a big guy, but he was fearless, and walking through the yard on his first day, Torres writes, he "saw eyes staring at him with sexual passion." Approached by a large, muscular inmate, Diamond was told not to worry about anything, that he'd be safe. But there was one condition: He'd have to be the inmate's wife. Diamond looked into the big man's eyes, then took the man's head in his hands, "and clamped his teeth on [his] ear, tearing at it in a rage." Screaming in agony, the inmate ran away. Diamond turned toward the other inmates who had been watching, "spat out the bloody flesh," then swaggered past them.

He was given a month in solitary confinement. Once again he remained true to the code of the streets and didn't tell the prison officials why he did it. After he was released from solitary, he took on all comers and soon established himself as the toughest guy in Green Haven, where he served eight years.

"Years later, after I had met Artie," Torres recalls, "I befriended [Green Haven's] warden and its chaplain. Out of these relationships came a parole for Artie," who went to work as Torres' conditioning coach.

For Torres, "having an ex-con near me and my wife worried me a little. Especially an ex-con of legendary proportions. But everything worked out, though I could never really figure Artie." Torres knew Diamond "was filled with violence," but "with my wife and me he was soft and gentle, almost weak."

Diamond soon became involved with people who "were planning to holdup a Bronx factory payroll," Torres writes. He was caught and sent back to Green Haven. He served his time. The years passed. He and Torres lost contact with each other. It was as if Artie "had vanished into the mist of time . . ." Until that night in 1976.

## *Long Ago*

In what I call another lifetime, I taught adult education classes in a Connecticut prison. Great job. The inmates were cooperative, especially the ones twenty-one and over. The guards? Well, two out of every fifteen were assholes. Shouldn't have been working with inmates. The rest were good people, bright, and hard workers.

Early one afternoon an inmate stopped me in the hallway, introduced himself, and said he was from Yonkers, New York. He was big and burly, probably in his late fifties. Wore a big Afro. He heard that in my class we had read an article about Artie Diamond. He said he served time with him. He said he wasn't in the yard that memorable morning when Diamond took a chunk out of the inmate's ear, but had watched the ex-fighter prove how tough he was many times.

"You'd have to kill the son of a bitch to stop him once he started throwing punches," he said. "He didn't like black people, but get yourself on his good side you were gold. Challenge him he'd hurt you. Real bad. I'll be transferred in probably two days to Somers, but I'd love to read the article. Bring back a lot of memories."

I said that was fine, and "I'll run you off a copy. Keep the article. I typed it up and mimeographed it and still have a bunch of copies. A relative who has been reading *The Sun Herald* for about a hundred years sent it to me. Said it'd make interesting reading. He was right."

*Yes, the old mimeograph machine. Better hold your breath while you filled it with fluid because one inhale gets you high for three days. That was probably why inmates of all shapes, colors, sizes, and crimes enthusiastically volunteered to run off copies of anything for me.*

*And I thought they wanted to be helpful.*

Later, I checked the inmate's rap sheet, the one who said he served time with Artie Diamond, and learned he wasn't bullshitting and did indeed do a stretch at Green Haven with Diamond.

*Requests from inmates come in all shapes and forms. The former Green Haven inmate's was sensible. But consider what a few weeks later, after a tall, skinny, large-eyed inmate sashayed into the classroom and sized it up, asked me.*

*"I'd like to sign up for class," he says.*

*Reasonable enough.* "Sure," *I reply.* "You have a few guys ahead of you, though. You'll have to wait a while. Maybe a few days."

*Before I could ask the inmate's name, he says,* "Do you offer a class in Sanskrit?"

*I wanted to say,* "Not this semester, young man, because we're offering advanced Latin and advanced hieroglyphics—but Sanskrit for beginners is on tap for next semester." *Instead I say something like,* "I'll get back to you next week and let you'll know exactly what we're offering next term."

# Uncle Cheech Talks Abeetz

October 30, 1979. I brought my battery operated tape recorder to Uncle Cheech's house on Kenny Drive, in East Haven, to interview him. We went outside—the day was brisk and sunny—and sat down at his very old and dangerous picnic table.

Dangerous? You bet. But Uncle Cheech would always remind me to check my *tuchis* for splinters before I left.

We began.

"Just talk? Well, okay. I'll begin with Pepe's Apizza on Wooster Street. Okay? The New Haven old-timers didn't call it apizza, you know. For them it's abeetz. I still call it abeetz. You got that? Abeetz. A-B-E-E-T-Z! Today most people call it pizza, right?

"Anyway, walk into Pepe's, take a deep breath, and you know you're in abeetz heaven. Pepe's serves nothing except abeetz. The ovens have been around since Frank Pepe opened the place in 1928. Back then he sold abeetz by the slice. Cost a nickel I was told. Thin crust. Expensive olive oil. He didn't skimp—not Mr. Pepe—and it paid off.

"There was one particular waitress there—middle-aged, definitely Italian, heavyset, shoulders like a football player—who almost always waited on me. (At least it always seemed she waited on me.) She'd shuffle to my table, and when she arrived I better know what I wanted. No dilly-dallying. And believe me I made sure I knew what I wanted. Either a small or medium mozzarella. With her tiny pencil in hand, she'd write my order on one of those lined waitress pads and then march off.

"And the way the abeetz was sliced. No two pieces were the same size. Careful though. It was hot! Grab plenty of napkins, too. I'm hungry talking about it.

"I liked baseball almost as much as boxing. The Phillies of 1950—the Whiz Kids—were my team. When they won the National League Pennant in 1950—they hadn't won one since 1914 or '15—I got so excited I almost barfed. Honest. I loved the entire team, especially Richie Ashburn. He saved the game that day against the Brooklyn Dodgers when he threw out Cal Abrams from centerfield in the last of the ninth inning with the score tied, 1-1. If Abrams had scored, the Dodgers would've won and forced a playoff. Then in the tenth, the top of the tenth, with two men on base, Dick Sisler, the great George Sisler's son, hit a homerun to left field. Then Robin Roberts—and what a great pitcher he was!—got the Dodgers out in order in the bottom of the tenth.

"And we won. We won the NL pennant!

"Of course, in the World Series—and I still can't figure out why it's called a *World* Series because it involves only American teams, right?—those Yankees won four straight. My heart broke. I could never figure out why Eddie Sawyer, the Phillies' manager, started relief pitcher Jim Konstanty in the opening game. Why didn't he start Roberts? Konstanty was the National League's best reliever, but didn't start a game all season. Appeared in 74 games in relief and was voted the National League MVP for '50. But Jim Konstanty wasn't a starting pitcher.

"My pick for the best heavyweight ever is Gene Tunney—which might surprise you. He could box and punch and move. And in his two fights against Dempsey, he was too fast and young for him. Dempsey's legs were gone. They fought twice, both ten round bouts, and Tunney won sixteen, maybe more, of the twenty rounds.

"After 1950 the best fighter pound-for-pound was Archie Moore, the Old Mongoose. Of course, you have Ray Robinson and Willie Pep and Joe Louis and Rocky Marciano and Muhammad Ali.

"Willie Pep had 241 fights and won 229 of them. A few years ago an aquaintance showed me how little he knew about boxing when he said Pep was overrated. Imagine! (I wasn't at all surprised that he took some mindless position in education and later developed an uncontrollable foot fetish.)

You know, supposedly Pep once won a round without ever throwing a punch. (Could be true.)

"And who knows how good Mike Tyson would've have been? His uppercut was deadly. Then there was Roberto Duran. As a lightweight he was unbeatable. Tommy Hearns—first as a welterweight and later a middleweight—was fierce. I'll never forget his knockout of Duran. And that destroyer—Marvelous Marvin Hagler.

"You know, and I've always enjoyed seeing this, when a fight is over fighters usually hug each other. Sometimes they even kiss. The same fighters who fought and hated each other, who wanted to hurt each other, often become close friends after they retire. Take Tony Zale and Rocky Graziano, Willie Pep and Sandy Saddler, Louis and Schmeling. It's a long list. Why? These guys realize they brought out the best in each other.

"Movies? I love *film noir*. Richard Conte, an Italian boy from New York, and McGraw—Charles McGraw—seemed to be in every *film noir* ever made. In many of Conte's movies he was usually returning from somewhere— maybe from jail or from the service. In *Cry of the City* he gave his best performance. McGraw's best movie was *The Narrow Margin*.

"Flynn, Garfield, Cooper, Lancaster—all terrific. The best American actors, the most talented, were Marlon Brando and Montgomery Clift. Case closed.

"Actresses? Ingrid Bergman was wonderful. Always earthy. Ava Gardner. Rita Hayworth. Gene Tierney. Bette Davis. They were all great. Oh yeah, there was Louise Brooks. Silent film star, though she made a few talkies in the thirties. She was one of a kind. She wrote *Lulu in Hollywood*. I haven't read it yet, but I will."

*After leaving Paramount Pictures in 1929, Louise Brooks, age twenty-two, journeyed to Berlin, Germany, at the request of director W.B. Pabst, to be cast as Lulu in* Pandora's Box. *In her book of autobiographical essays,* Lulu in Hollywood, *Brooks learned from Pabst that actors had a "natural animosity" toward other actors, "present, living or dead." Pabst encouraged that hatred among cast members. He believed it "preserved their energy for the camera."*

*Brooks wrote that Fritz Kortner, who played Dr. Schon and whom Lulu murdered in* Pandora's Box, *disliked her because he felt she couldn't act and had cast some weird spell over Pabst. (Brooks calls it a "mysterious alliance . . . a kind of wordless communication." Lotte H. Eisner, in "A*

Witness Speaks," calls it "a somewhat magical force emanating from this strange girl . . .")

Pabst used Kortner's "true feelings" about Brooks in a scene that gave "him the opportunity to shake me with such violence that he left ten black-and-blue finger prints on my arms."

In Hollywood, before working with Pabst, Brooks admitted she was just "a pretty flibbertigibbet," but under Pabst's direction, in both Pandora's Box and The Diary of a Lost Girl (1929), "I became an actress. I would be treated by him with a kind of decency and respect unknown in Hollywood."

The hatred Pabst encouraged between actors is applicable to prizefighters. When the bell rings, you and your opponent hate each other. You both set out "to hurt each other," Mike Tyson says. "It's brutal. But that's boxing."

Consider Roberto Duran. In Spring Toledo's The Gods of War, he says that Duran, when he was at his best, "was something like preternatural malevolence." For his first fight with Sugar Ray Leonard, in 1980, fought at Olympic Stadium, Montreal, Canada, Duran's trainers, Ray Arcel and Freddie Brown, "closed the blinds and applied old school methods in the shadow of Stillman's Gym. They brought a Panamanian to a peak of human performance so perfect in its blend of science and ferocity that it would never be approached again—by Duran or anyone else."

Over time the once fierce Duran become "a melting legend . . . As the curtain slowly descended on a career that spanned five decades, there was little left that recalled what was; just some old tricks in an arsenal ransacked by age and an unbecoming appetite."

"Anyway, if you want a contemporary actor I admire, it's Meryl Streep. I first saw her in a TV movie, *The Deadliest Season*. That was in 1977. I knew right away she was special. She played Sharon Miller, the wife of Gerry Miller, a hockey player charged with manslaughter. Michael Moriarty played Miller. He had attacked an opposing player during a game, but he was cleared of any crime.

"The movie ended with a long shot of Miller skating around the rink after a practice. Alone. Going nowhere. Skating in circles."

# Cut and Shoot

*"Try hard as I might, I could not remember what I had dreamed."*
—Neil Gaiman

It started with a dream. On a trip to New Haven several weeks after the dream, I was driving on Whalley Avenue, approaching Norton Street, and on my right was where the Whalley Theater was once located.

You see, I grew up in the Elm City and the Whalley Theater was almost a second home. I saw my first movie there, Claude Rains in *The Phantom of the Opera*. Never forgot it. That was where I saw *The Adventures of Robin Hood*, too, and formed a lifelong admiration for Errol Flynn. (Don't believe all the negative crap you read about him.)

Then, like that bolt of lightning from the sky, the one that changed young Billy Batson into Captain Marvel, I remembered in my dream heavyweight champion Floyd Patterson was defending his title against fifth ranked Roy Harris of—take note— Cut and Shoot, Texas. It was screened at the Whalley Theater, and Harris pulled off a tremendous upset by knocking out Patterson in four rounds: A right cross followed by a left hook put Patterson down for the count. Of course, in real life Patterson stopped Harris in the 12$^{th}$ session, and the fight wasn't screened at the Whalley but at the New Haven Arena. Patterson was dropped in the second round but quickly got back to his feet. For Patterson, the knockdown was a wake-up call. From that point on, Red Smith writes, Harris "got his face punched off for $100,000 . . . After Floyd Patterson had knocked the challenger down four times in twelve rounds with blows that sprayed blood over the ringside rows, Mushy

Callahan, the referee, [stopped the fight]. It went into the records as a twelfth round knockout. It will go into memory as the best fight of Patterson's young life against the gamest and strongest adversary he has ever met."

In *Once They Heard the Cheers,* W.C. Heinz devoted a chapter to the fighter from Cut and Shoot who graduated from Sam Houston State Teachers College in Huntsville, Texas. After he retired he taught in an elementary school in Huntsville.

Nineteen years later Heinz returned to Conroe, Texas, to its County Court House, where Harris worked as County Clerk, to interview him for the second time. Harris told Heinz that he returned home, studied law for three years, passed the bar exam and became a lawyer.

Early on in their interview Heinz told Harris he had recently talked with Floyd Patterson. Harris asked about Patterson, how he was feeling, and Heinz said that he was fine but upset about something Harris said about him.

Harris said he liked Floyd and was sorry he was upset at him. Turning psychiatrist, Harris told Heinz that Patterson had an inferiority complex and didn't want to hurt an opponent until the opponent hurt him.

Heinz said that for a fighter Patterson was too kind.

Later Harris went on a rant, which Heinz quotes: America is "being infiltrated by other countries tryin' to tear us down . . . They find a minority group, and they shove and agitate them and push 'em, and everybody feels a little sorry for himself anyway. . ." Infiltrators persuade Native-Americans "how bad they were treated, and the Nigra they tell how he's been mistreated. Every minority they sell a bill of goods. You can't live in the past, and I don't think our government has put enough emphasis on braggin' on our country. Our country is the greatest country in the world. We've got the greatest people in the world."

The planet! The planet! Which one are you on, Mr. Harris? *Braggin'* is another word for self-righteousness, and self-righteousness is negative and nonproductive. Face it, Mr. Harris, our country is imperfect, and the way to improve it is to admit its many imperfections, then correct them—because they're correctible. It takes hard work, a change of attitude, not denial about the country's many flaws.

# The Milling Art

*"But a fellow who came up at two was so strong he would bear investigation."*
—A.J. Liebling

Though it's rightly considered a classic of boxing journalism, A. J. Liebling's *The Sweet Science* is, says Joyce Carol Oates in her bold and insightful *On Boxing*, "a peculiarly self-conscious assemblage of pieces, arch, broad in its humor, rather like situation comedies in which boxers are 'characters' depicted for our amusement . . . he is pitiless when writing about 'Hurricane' Jackson, a black boxer cruelly called an animal, an 'it,' because of his mental inferiority."

Oates is right about Liebling's insensitive and unfair assessment of Jackson, a heavyweight with a windmill style who came from Rockaway Beach, New York, and fought from 1951 until 1961.

The day before Jackson fought Nino Valdes at Madison Square Garden, in 1954, Libeling writes, "I had nothing against Jackson *qua* Jackson . . . but if the animal could beat even a fair fighter [like Valdes] it meant that two hundred and fifty years of painfully acquired experience had been lost to the human race; science was a washout and art a vanity . . ."

Referee Al Berl stopped the fight after Valdes, several inches taller than Jackson and twenty pounds heavier, with "shoulders [that] look as wide as a door," floored Jackson for the third time in the second round. In 1954 the rule in New York was that the referee must stop a fight when a fighter was knocked down three times in one round. Though well intentioned, Liebling writes the rule was "silly, because a boxer like Jackson, who doesn't know what to do with his feet, can be knocked down several times without being hurt much, while a fellow who is helpless but remains upright takes a beating without respite, the kind that is most likely to end in permanent injury." Maybe.

In contrast to Liebling's unkind words about Jackson, his descriptions of the battles between Rocky Marciano and Jersey Joe Walcott, Archie Moore and Marciano, and Sugar Ray Robinson and Joey Maxim are spot-on.

The year was 1952, the place Philadelphia's Municipal Stadium, and Libeling says the right hand Marciano knocked out heavyweight champ Joe Walcott with was "a model of pugilistic concision [as] both men started right leads for the head at the same moment." Slightly ahead on the judges' scorecards, Walcott had been punching faster and sharper and "figured to get in there first in such an exchange." He was wrong: Marciano became stronger as the fight progressed, his punches more accurate, and "according to old-timers about as hard as anybody hit anybody."

An example of Liebling at his most specific and vivid is his wonderful description of Marciano's knockout of Walcott, who "flowed down like a sack of flour out of a chute. He didn't seem to have a bone in his body . . . Rocky knocked him out with the kind of punch he wasn't supposed to know how to use."

Moore's second round knockdown of Marciano, in their 1955 title bout at Yankee Stadium, came from a left hook thrown "with classic brevity and conciseness, and Marciano went down—for two seconds . . . A man who took nine to come up after a punch like that would be doing well, and the correct tactic would be to go straight in and finish him. But a fellow who came up on two was so strong he would bear investigation." In the ninth round, Liebling continues, "Marciano slugged him down, and he was counted out with his left arm hooked over the middle strand as he tried to rise." Moore's courageous bid for the heavyweight title was over.

Marciano retired after his victory. Moore continued fighting until 1963.

In 1952, at New York's Yankee Stadium, Sugar Ray Robinson's right hand shot to light heavyweight Joey Maxim's jaw sent the bigger, stronger Maxim reeling backward but not down. Instead, Maxim, far behind on all the officials' scorecards, kept fighting "in his somnambulistic way"—and that's when Robinson knew he couldn't score a knockout. At the same time, he knew that if he remained upright until the final bell, he'd win the decision. But Robinson had no strength left, and when the bell sounded for the thirteenth round, he was finished. "He had, quite simply, collapsed from the heat, like a marathon runner on a hot day."

Beaten by his ego, the heat, and Maxim, there would be no fourth world title for Sugar Ray Robinson, perhaps the greatest fighter of all time.

# Toledo Slaughter

***Beaten***
A fighter isn't ever the same after killing an opponent in the ring. Maybe a piece of him dies when his opponent does.

For heavyweight champion Jess Willard to beat challenger Jack Dempsey, he needed the killer instinct—which Dempsey possessed in abundance. Willard didn't possess it. Maybe he never did.

It's possible that Willard, who on more than one occasion told reporters he didn't like hurting people, was a beaten man before he fought Dempsey: Willard was involved in a tragedy six years prior, in 1913, when he fought a young Wyoming heavyweight named John "Bull" Young. In the eleventh round, "[Willard's punch] drove the base of Young's jaw into his brain," Roger Kahn writes in *A Flame of Pure Fire: Jack Dempsey and the Roaring 20s*. Young died the next day.

Thirty-one years later Max Baer, like Willard, killed his opponent in the ring. The fighter was Frankie Campbell, a highly regarded prospect from San Francisco, California. "Baer trapped Campbell in a corner [in the fifth round] and landed a tremendous left hook to the head," writes John Jarrett in *Dynamite Gloves: The Fighting Lives of Boxing's Big Punchers*. Campbell stumbled into a corner but didn't go down, as "Baer continued to slug him with crushing punches, finally bringing him down with that right hand hammer."

Campbell was taken to Mission Emergency Hospital, "where he died the next morning of a cerebral hemorrhage." Baer was charged with manslaughter.

Jarrett quotes Baer, who told Nat Fleischer, founder and editor of *Ring* magazine, that "It was a tremendous relief when the surgeons announced that Frankie had died of a brain concussion and the court ruled it was an accident and cleared me of the manslaughter charge. But for the time being, I felt as if I never wanted to see a boxing glove or enter a ring. My enthusiasm for the game had gone." (Some boxing historians believe Baer never regained his passion for the sport.)

## *A Whistle*

Toledo, Ohio. 1919. A hot, cloudless day and 50,000-plus fans crowded into the Bay View Park Arena to witness Jack Dempsey fight Jess Willard, who became heavyweight champ in 1915 when he stopped Jack Johnson in Havana, Cuba, in twenty-six rounds. A year later he successfully defended his title by knocking out Frank Moran in eleven rounds. Nicknamed "The Pottawatomie Giant," at 6'7" he was six inches taller than Dempsey and weighed 250 pounds, almost sixty pounds more than the young challenger.

According to Randy Roberts, in *Jack Dempsey: The Manassa Mauler*, after Willard entered the ring, Dempsey saw him and said he was fighting not only for the title but also for his life.

This happened: In round one both fighters fought the opening minute cautiously. Then Dempsey fired off "a five punch combination," Roberts writes. " . . . the first four blows—two lefts and two rights—landed solidly to Willard's body; the last punch—a devastating left hook—connected with the right side of the champion's face. Willard went down. As soon as Willard stood upright, Dempsey was upon him aiming every punch to the champion's head. Pinned against the ropes, Willard was helpless."

A left hook sent Willard down for the second time. "Only with the aid of the ropes was he able to stand, and while he held on there, turning away from the challenger, Dempsey hit him from behind with more punches. Another left hook sent the champion down for the third time. Veteran referee Ollie Pecord didn't prevent Dempsey from circling behind Willard. This was one of Dempsey's favorite tactics: whenever he knocked down an opponent, he tried to get behind the unfortunate foe so that his next punches would be unseen by the victim."

After Willard went down for the seventh time in the round, it appeared he had been counted out. After being announced as the new world heavyweight champion, Dempsey left the ring. But it was soon discovered that before Referee Pecord reached ten, the whistle, inaudible to the officials, had been blown ending the round. (That's right! A whistle, not a bell.)

Dempsey was almost in his dressing room when his manager Jack Kearns caught up with him and said the fight wasn't over. So back to the ring marched Dempsey and the fight—the slaughter, really—continued for two more rounds. Willard was unable to come out for the fourth session.

"Today if there were a fight such as this," Roger Kahn writes, it would have ended "after the third knockdown. A good referee would simply stop the fight. Failing that, a good doctor would *demand* a halt . . . Ollie Pecord did nothing of the sort."

### Loaded

There were rumors that Dempsey's gloves were loaded. But according to Randy Roberts, "Jack Kearns' second son, Jerry McKernan, told the story of how [his father said he] put plaster of paris on the bandages covering Dempsey's hands before the Toledo fight." But Jimmy DeForrest, the man who taped Dempsey's gloves, called the story groundless, and that he used "a hard but perfectly legal adhesive tape."

According to Roberts, noted sports journalist Dan Daniel, who was at ringside, "bluntly claimed that one could not believe anything Kearns said. Later, an attempt to duplicate the method Kearns said he used proved to be an utter failure. The plaster of paris did not dry in the prescribed time and it crumbled as soon as it hit another object."

# Gentleman Jack

*Dempsey vs. Gallico*

Participatory journalist George Plimpton's purpose in sparring with light heavyweight champion Archie Moore in 1959, at New York's Stillman's Gym, was to make the experience the centerpiece of *Shadow Box: An Amateur in the Ring*, his singular book about fighters, trainers, and writers who penned books and articles about them.

Maybe Plimpton got the idea of sparring with a professional prizefighter from sports journalist Paul Gallico. In 1923, Gallico climbed into the ring against heavyweight champion Jack Dempsey, who was training for his title defense against Luis Firpo. His purpose was, Randy Roberts writes in *Jack Dempsey*, to experience firsthand "just how a knockout feels," and then write about it for the New York *Daily News*.

Enter the paranoid Jack Kearns, Dempsey's manager, whom Roberts says was certain that Gallico, standing six feet three inches and weighing 190 pounds, "wasn't a reporter but someone sent by the Firpo camp to give the champion a tough time." Soften him up. Embarrass him. "Although Kearns was unable to convince Dempsey to cancel the bout, the manager did arouse suspicions in the champion's own mind." (Not good news for Gallico.)

The sun was shining when Gallico entered the ring. It was still shining when the bell rang, Roberts writes, "and Gallico assumed 'Pose A' from the *Boxer's Manual*." He landed several soft jabs to Dempsey's face and was beginning to enjoy himself. Then reality struck. The sun disappeared. Everything turned dark. Gallico found himself on the canvas. After he regained his footing, Dempsey—who before the fight "seemed irritated with [Gallico's] presence"—gave him "a half-dozen affectionate taps on the back of the neck." Affectionate? Gallico went down again—and stayed down.

Gallico emerged from the session with a bloody nose and a pounding headache. "[His] only consolation was that he received a by-line for his courage against the champion," Roberts says.

### *Three Years*

Talk about a wild, surreal fight! Take Dempsey's successful title defense against Luis Firpo, the "Wild Bull of the Pampas." It produced eleven knockdowns and ended in the second round. (Dempsey was floored twice, Firpo nine times.)

After the Firpo fight Dempsey was inactive for three years, a lifetime in professional boxing. He performed on the vaudeville circuit, trained irregularly, fought occasional exhibition bouts, met with President Calvin Coolidge, was operated on for hemorrhoids, had his broken nose "recast along Grecian lines," writes Roberts, and enjoyed New York City's night life. Then came Hollywood. There, Universal Studios cast him in several terrible but popular movies. In 1925 he married actress Estelle Taylor, whom he divorced five years later.

Before losing to Gene Tunney in 1926, at Sesquicentennial Stadium, Philadelphia, Dempsey had defended his heavyweight championship four times. His opponents were Billy Miske, who was terminally ill; Georges Carpentier, who was tough and skilled but much smaller than Dempsey; Tommy Gibbons, who was past his prime but went the distance because, like Tunney three years later, his boxing skills enabled him to withstand Dempsey's rushes; and Firpo, who knocked Dempsey out of the ring and should've been declared heavyweight champion had not ringside reporters pushed the champion back into the ring before the count reached ten.

Excuses were made for Dempsey's first loss to Tunney. Writer Ring Lardner claimed the fight was fixed. (That was because he bet on Dempsey.) There were rumors—never proved—that Dempsey was poisoned. None of those excuses is credible. Dempsey was defeated because Tunney was the better man! Younger than the champion, in superb condition, Tunney was in his prime, while age had caught up with Dempsey and his absence from the prize ring had dimmed his reflexes and timing.

## Meeting Dempsey

Grantland Rice had mixed feelings about Dempsey as a fighter. But after Dempsey retired he said Dempsey was one of the finest men he ever met. A real gentleman.

At first, Roberts also found it "difficult to believe that the Dempsey who fought desperate men in saloons and pounded [Jess] Willard's face into a bloody mask was a gentleman. As a contender and champion Jack Dempsey was hungry. Inside the ring he moved with restless, violent energy. And for most of his career, he was sullen and brutal outside the ring; the gentleman inside rarely struggled to get free."

Roberts learned that in retirement Dempsey's *internal gentleman* had emerged: He was generous with his money, which he accumulated by training boxers, refereeing wrestling matches, and becoming the owner of a restaurant in New York.

"Paul Gallico estimated that Dempsey in his lifetime gave a fifth of everything he earned as gifts and handouts to old fighters whose handlers had abandoned them to poverty," adds Roger Kahn, in *A Flame of Pure Fire*.

The first and only time Roberts met the former champion was arranged by Dempsey's adopted daughter. The meeting took place on December 26, 1973, at the former champion's restaurant.

A graduate student at the time, Roberts recalls, "My coat was too thin, my hair too long, and I had not enough money."

But nothing was going to stop Roberts from meeting the legendary champion. And when they did meet, Dempsey "gracefully smoothed awkward introductions. And he did much more than simply permit me to interview him. He invited me to join him at dinner with some of his family and friends. This was more than I dared to expect . . . Dempsey was a gentleman and a perfect host, trying to bring out the best in his guest. He was concerned with the feelings and desires of the people at the table—so concerned, in fact, that at times . . . I found myself doing the talking and Dempsey doing the listening . . . It was a memorable evening with a memorable man . . . Perhaps Grantland Rice was correct. It took a remarkably gentle man to live Dempsey's post-championship life as decently as he lived it."

# Dempsey According to Roger Kahn

***Welcome to Harding's White House***

For some readers Roger Kahn's persuasive *A Flame of Pure Fire* is too long, too digressive. Not for me, though. True, Kahn concentrates mostly on Jack Dempsey, though he wrote about other noteworthy figures of the twenties, like President Warren G. Harding and Henry Ford who were, he says, "an integral part of that era and it would be unbridled romanticism to say Dempsey all by himself symbolized the Roaring Twenties."

The handsome Harding "was a well-established philanderer," and "you had to be careful where you went in Harding's White House . . . You never knew behind which door you might find a naked president giving the best of himself to an equally naked young thing."

Kahn quotes Frederick Lewis Allen, in *Only Yesterday*, who sums up Harding as "A common small-town man, an average sensual man . . . His private life was one of cheap sex episodes . . . Even as President he was supporting an illegitimate child born hardly a year before his election . . . An ambitious wife had tailored and groomed him into outward respectability . . . yet even after he reached the White House, the rowdies of the Ohio gang were fundamentally his sort; it was in smoke-filled rooms that he was really most at home."

As for the anti-Semitic Ford, he published a vitriolic weekly newspaper, the *Dearborn Independent*. He believed that "the Jews were plotting to dominate the world. At Ford's direction, editors at the paper depicted Jews as a monochromatic group, united behind the single cause of domination. Jews already dominated finance. Jews had already created pornography to corrupt and weaken Christian youth . . . Not only had Jews murdered Jesus Christ, Henry Ford preached, they were murdering Christian infants for their blood."

Kahn correctly replies, though much too kindly, "It was not Jews who were out to dominate the world. The people who wanted domination were the anti-Semites."

## *An Angry Rice*

Challenger Jack Dempsey and heavyweight champion Jess Willard fought on July 4, 1919, in Toledo, Ohio. While most of country was excited about the bout, sports journalist Grantland Rice was angry: For Rice, the fight was *only* a sporting event.

"As I got the story in a later time," Kahn writes, "Rice spoke with Ring Lardner along these lines: 'I covered the Battle of Argonne Forest [in northeastern France]. Forty thousand American soldiers died there and only forty newspapermen covered that slaughter, which is as much as I hope I ever have to see of hell. Now we have a couple of brawlers getting ready for a fistfight and we have six hundred reporters cheering them on like heroes here in Toledo. It doesn't make sense . . . These guys can punch, all right, but they didn't have the guts to fight for their own county.'"

In 1920, Dempsey went on trial for draft evasion. He was found not guilty, but Rice, along with the *Herald Tribune*'s Bill McGeehan, continued denigrating him in print. (They didn't criticize Babe Ruth, Benny Leonard, Red Grange, and Bill Tilden who, like Dempsey, didn't serve in the military during WWI.)

Then came WWII. Kahn writes: Told that he was too old to enlist in the Army, thanks to "some high-level intervention, perhaps by Franklin D. Roosevelt, the Coast Guard accepted Dempsey, commissioned him, and put him in charge of physical training at its biggest base, Manhattan Beach . . . In time he was promoted to lieutenant commander . . . one of Dempsey's units was assigned to move in with the landing parties storming Okinawa. As the young warriors climbed into small boats for the assault on April 1, 1945, a line officer said, 'You stay here with me, Jack. We can't afford to lose you' . . . Dempsey said, 'Sir, I trained these boys and they look up to me. I go where they go.' Which is how Jack Dempsey hit the beach at Okinawa when he was forty-nine years old."

## *We See What We Want to See*

In *Gene Tunney: The Golden Guy Who Licked Jack Dempsey Twice*, John Jarrett writes: While training for Dempsey, Tunney "would run backwards for miles at a time, as much as some fighters would run forwards, having learned from medical studies that backwards running really gave the cardiovascular system a tough workout."

Running backwards paid off. In round seven of their second fight—maybe the most controversial round in boxing history—Tunney, back on his feet after being knocked down, began backpedaling. Dempsey gave chase but couldn't catch him. Tunney's young legs were too strong for Dempsey's old ones.

In round eight Tunney floored Dempsey with a combination to the head. "Dave Barry sprang at Dempsey and shouted, 'One!'" Roger Kahn writes. "Tunney had not gone to the farthest neutral corner. He was hovering nearby, as Dempsey had done the round before. Barry ignored him. Dempsey was back on his feet before the count reached two."

Kahn is outraged: "Two knockdowns, one round apart, and two different sets of rules. The explanation is not complicated. In my tape of Chicago 1927, I am looking at a crooked referee."

Was Kahn objective in his assessment of the fight? Judge for yourself. He knew Dempsey personally and greatly admired him. For him, Dempsey "is a rare thing: close up, as from afar, he was a hero."

Consider the Dempsey-Luis Firpo fight, probably the most exciting three minutes and fifty-seven seconds in boxing history. Early in round one the fighters clinched. As referee Jack Gallagher broke them, Jack Cavanaugh writes in *Tunney*, "Firpo dropped his arms while stepping back in accordance with Gallagher's instructions. It was a huge mistake. Dempsey immediately ripped across a left uppercut to the head that sent Firpo reeling to the canvas for a nine count."

Kahn sees it this way: The fighters clinched and "the referee ordered, 'Break!' As they did, Dempsey drove an uppercut into Firpo's chin. Firpo fell backward awkwardly, as if unconscious." No warning from referee Gallagher! Not a hint of Gallagher being crooked from Kahn!

# Don't Crowd the Champ

*I'm There*

With little effort I made time flip back to 1923—and there I was at world heavyweight champion Jack Dempsey's training camp at Sulphur Springs, New York. His title defense against Luis Firpo at the Polo Grounds, in New York, was a few weeks away, and he had just finished sparring with New York *Daily News* sports reporter Paul Gallico, whose assignment was to pen a first-hand account of what it felt like to be hit by Dempsey. (Why not have a friend hit you on the head ten times with a hammer, Paul?)

Gallico was, you see, certain the champ would take it easy with him, but what he didn't know, says Jack Cavanaugh in *Tunney*, was that Dempsey believed "anyone who got into the ring against him was a mortal enemy."

The session lasted less than a round. Gallico came out, flicked out a few harmless jabs to Dempsey's kisser, and found himself on his rear end. Reality had struck! Gallico somehow scrambled back to his feet.

"Hang on to me, kid," Dempsey advised. "You'll be okay. Your head'll clear. Just hang on." Then Dempsey whacked Gallico on the back of his neck and down he went again. This time he stayed down. (Smart thinking.) When the session was over he had a swollen lip, bloody nose, colossal headache, and a story with a byline.

There were, of course, celebrities who wanted to spar with Dempsey. Something to brag about, I guess. Sometimes Dempsey would oblige. But Ernest Hemingway, who considered himself a fairly good semi-professional pugilist, was one celebrity Dempsey avoided. At the time Hemingway had been writing poems, Roger Kahn writes in *A*

*Flame of Pure Fire* that "were not very good and short stories that were extraordinary."

So here's the great Dempsey, who hasn't worked up a sweat against Gallico, sitting across from me at a bench several yards from the boxing ring where he and Gallico have sparred. (And I didn't have splinters to worry about.)

I begin: "Champ, you've sparred with many non-boxers— celebrities— but you made it a point to stay away from Ernest Hemingway who once challenged you." (No sense wasting time, right?)

"Yes," Dempsey says, looking me in the eye and nodding. His expression altered slightly. I began feeling uncomfortable.

"W-why, c-champ?"

"I'd hurt him. Really hurt him. Real bad!" Dempsey paused. "Move onto another subject, okay?"

I hope the champion didn't hear my gulp.

I ask him if there was anyone, any fighter, he was afraid of, and without missing a beat he replies, "Sam Langford. It wasn't Harry Wills. It was Langford. I could've taken Wills. He was slow. Langford was smaller than me—I think he was five foot seven or eight—but he could hit and he was fast. Yeah, I was scared of Langford. He would've beaten me."

**I'm Back**

Again, with little effort I made time flip and was back in 2015, at home, in Preston, CT. That night I dreamed I sparred with Dempsey. Dreamed that he complimented me on my snappy left jab and dazzling foot work. Dreamed that he envied my incredible hand speed and conditioning. Dreamed that he marveled at my gentility.

After several cups of black coffee the next morning, I went to the local library to research what most likely caused the animosity between Dempsey and Hemingway, found Roger Kahn's book about Dempsey, and learned what I needed to know, thank you.

"Probably with a few drinks in him," Kahn writes, "Hemingway challenged Dempsey in Paris, 1922. Dempsey recognized the mad brutality that was as much a part of Hemingway as writing genius and declined. I can imagine Dempsey dismissing boxing Hemingway with an easy putdown, and Hemingway persisting and Dempsey, who is now annoyed, sneering or perhaps even laughing in Hemingway's face.

On certain matters it was not a good idea to crowd the champion. Any amateur who threw down a serious challenge was delivering an insult and it is remarkable that Dempsey remained as gentle as he did with such pretenders."

# Stanley the Man

*Murdered*

In *The Killings of Stanley Ketchel,* James Carlos Blake says that Ketchel, born Stanislaus Kaicel and nicknamed the Michigan Assassin, "knew what he loved most about fighting was its clarity . . . when you knock a man out you resolve matters with an absoluteness impossible to rhetorical arguments or philosophical disputes."

It was in Conway, Missouri. October 10, 1910, that Ketchel, world middleweight titleholder from 1907 to 1909, was living and working at the farm of Colonel H.P. Dickerson, a family friend. While Ketchel was eating breakfast, Walter Hurtz, a farm hand, whose real name was Walter Dibley, shot him in the back. Hurtz had discovered that his common law wife, Goldie Hurtz, "widely known to have led a sordid personal life," had been sleeping with Ketchel.

Ketchel, age twenty-four, died at 7:03 that evening, at a Springfield, Missouri, hospital.

Blake writes that when news of Ketchel's death reached longtime rival Billy Papke, whom Ketchel had beaten in three of their four no-holds-barred fights, it "nearly broke [his] heart. Now he could never prove to the world he was the better of that son of a bitch."

When heavyweight champion Jack Johnson, Ketchel's conqueror, learned of Ketchel's killing, he paused and "looked off to seaward . . ." Then Blake quotes Johnson who, flashing a huge smile, says, "Dollar to a doughnut Mr. Stanley was starting to turn around to try and catch that bullet with his *teeth* . . ."

## Dreams

The dream motif gives Blake's vividly written, inspired, and often wild novel unity. In one dream, for instance, Ketchel is "fighting [heavyweight champion] Jack Johnson . . ." He throws a right hand and "didn't wake . . . until the punch struck. Struck and felt . . . *perfect* . . . and Johnson fell."

In the actual fight both combatants agreed to hold back. Fight to a draw. A few months later they'd fight again for real. Blake has Willus Britt, Ketchel's manager, predicting that "the whole country'll go crazy. They'll be screaming for a rematch. And that's when we make a real killing"

So a deal was made: Ketchel and Johnson would fight twenty rounds, no decision would be rendered, and Johnson would retain his title. But Ketchel and Britt had other ideas.

The battle takes place at Mission Street Arena, in Colma, California, in 1909. In the twelfth round Ketchel drops Johnson—who outweighs him by thirty-five pounds and is five inches taller—with a vicious right to the temple. When Johnson is back on his feet, Ketchel "leaps headlong, swinging from the hip," Blake writes. "It is a rankly reckless move and Johnson has seen it coming. And he beats Ketchel to the punch with a right cross that by his own admission is one of the hardest blows of his life."

Down goes Ketchel "like a bundle of the latest edition hitting the sidewalk at a newsstand and Johnson's forward momentum carries him into Ketchel's upslung legs and he trips and goes sprawling and immediately scrambles to his feet . . ."

*In his critically acclaimed* Unforgivable Blackness: The Rise and Fall of Jack Johnson, *Geoffrey C. Ward writes about the deal Ketchel's manager made with Johnson. For Britt, who knew his fighter didn't stand a chance against Johnson, money was the big concern: "Johnson and Ketchel were to split 40 percent of the proceeds for the films of the fight. The longer it went . . . the bigger the film's box office would be . . . Johnson agreed to let Ketchel make a good showing for at least twelve of the twenty scheduled rounds"*

*In the twelfth round, as we know, "Ketchel strayed from the script. He and [his manager], crouching below his corner, had evidently planned a double cross. Ketchel's right hand landed and down went the champion . . .*

*[Johnson] caught himself with one hand . . . sprang up again . . . a sheepish grin on his face."*

Johnson's right cross ended the fight. Ketchel was decked. In Ward's perfect word picture, Ketchel "stayed there, flat on his back, arms outflung. Four of his teeth were strewn across the canvas. The grim-faced crowd fell silent as the referee counted him out. Johnson stood, one arm along the top rope, the other on his hip, peering down anxiously for some sign that his opponent would revive. At least five minutes elapsed before Ketchel moved."

In *Dynamite Gloves,* John Jarrett quotes Johnson as saying, "I was more enraged than I was hurt. I don't think I ever hated two people as much as I did Willus Britt and Stanley Ketchel at that particular moment. They tried to double-cross me and steal my championship. I bounced back to my feet with murder in my heart."

Enter George Little, one of the many managers Johnson had hired and fired. Little's take on the Ketchel-Johnson fight, though it strains credibility, was that, Ward writes, "Everything, including the controversial climax, had been prearranged."

Little claimed he had been at Britt's house a few weeks before the fight watching Johnson and Ketchel rehearse the final round. Johnson was *"falling again and again with a suitably agonized look, then springing back to throw his uppercut."* But Ketchel protested, saying that he wasn't an actor! There's no way he could fake it! Johnson must really belt him!

## Many-sided

Blake's unforgettable depiction of Ketchel is believable and complex, not one-note. As a teenager he rode the rods, fighting anyone who challenged him. Size didn't matter. Fists, knives, and clubs didn't either.

And as a prizefighter fighter he was always, Blake writes, "near to insane in his ferocity." Consider his first professional fight on May 5, 1902, against Killer Kid Tracy at the Broadway Theater in Butte, Montana.

Ketchel weighed 143 pounds; Tracy 163.

In ring center the referee gave each fighter "the standard prefight warning against illegal tactics and then asked if they had any questions." Blake writes that Tracy glared at Ketchel and says, "Yeah, I got a question. Where you want the body sent?"

The bell rang. Ketchel rushed at Tracy and "during the next few seconds [Tracy] was hit so many times so fast that his head whipped from side to side as if vehemently denying all notion of continuing in this ill-chosen occupation. His mouthpiece sailed out of the ring . . . The Killer Kid lay spread-eagled and the ref didn't bother with a count . . . The fight lasted nine seconds."

When his beloved girlfriend Kate Morgan dies, we glimpse another side of this brash and dangerous, womanizing and bigoted individual. In some of his strongest prose, Blake writes that Ketchel—so fierce in the ring—"sat by the bed and held [Kate's] hand until it was cold in a way no living hand could ever be. In the company of her corpse he felt more alone than he'd known it was possible to feel. Felt as if she'd killed something in his own heart as surely as she had stopped hers."

# The Proper Pugilist

*Cashel Byron*

Well-researched and gracefully written, Jay R. Tunney's book, *The Prizefighter and the Playwright*, is about the relationship between heavyweight champion Gene Tunney, the author's father, and playwright George Bernard Shaw. It began after Tunney made a remark about Shaw's novel, *Cashel Byron's Profession*, which is about a professional prizefighter who closely resembled Gene Tunney in temperament and boxing style. According to Jay R. Tunney, his father said if Shaw's book were ever adapted for the big screen, "the story would have to be changed somewhat . . ." He stressed that he wasn't criticizing "the book's literary merit."

To the surprise of many British reporters, Shaw's response was "courtly" and he expressed the desire to meet the boxer. Shaw had been interested in boxing, which he regarded as a fascinating coupling of mental and physical skills. (There were rumors that in 1883 he boxed as an amateur in England.)

*Complete Sentences*

Sports journalists and most of the boxing crowd regarded Tunney as a snob because he read books, quoted passages from Shakespeare, and memorized poems by the great English romantic poets Keats, Byron, and Shelley. Also, this poor boy from Greenwich Village, New York, spoke in complete sentences and—imagine!—had the impudence to admit he wanted to be *more* than a professional boxer. Maybe a writer, or a businessman. Maybe both.

One of Tunney's many detractors was Will Rogers. Jay R. Tunney quotes him as saying, "Tunney just lectured before Yale's class on

Shakespeare. He said he read Shakespeare ten times before he could get what he meant. Now that brings up the question: Is there something wrong with Shakespeare or with Gene Tunney? If everybody has to read his stuff ten times, why, Shakespeare is not the author he is cracked up to be. But if somebody else can read him and get him the first time, why, Tunney is not the high-brow that *he* is cracked up to be."

Paul Gallico of the New York *Daily News* and Westbrook Pegler of the *Chicago Tribune* were two other critics who, according to John Jarrett in his book *Gene Tunney,* "freely admitted they sat up nights thinking of ways to belittle the boxer who preferred books to broads and booze."

Tunney went on the offensive. In *Tunney,* Jack Cavanaugh quotes the fighter as saying: "Being a boxer, my reading attracts attention. Some think I am high-hatting the boys when I talk about literature. I am not. I didn't begin to talk about books. Others did it to make conversation. Many put me through a cross examination . . . as if I were an infant prodigy of eleven applying for admittance to college. I answer out of politeness' sake . . . Because a man is a boxer it doesn't follow that he has to be illiterate."

To those sports journalists who "suggested he was a literary 'poseur,' and even a 'phony highbrow,'" Tunney pointed out that his interest in reading went back to his days at St. Veronica's School in Greenwich Village, where he learned the soliloquies of Shakespeare's Hamlet, Antonio, Portia, and Shylock.

Jay. R. Tunney quotes Yale Professor William Lyon Phelps who refutes Tunney's critics: "When I invited him to lecture on Shakespeare at Yale, I knew he would tell us not what somebody else thought of Shakespeare, but what, in daily life, Shakespeare meant to him. Tunney has an interesting mind."

*After Tunney retired, many of his critics acknowledged that his reading and preparing for opponents by studying their boxing styles and disciplining himself to live a clean life, both in and out of the ring, are characteristics to be praised, not ridiculed.*

*In his book* Farewell to Sport, *Gallico admits he and other sports journalists were wrong about Tunney. "He was caught groping for light, serenity, and education and ridiculed for it. He probably never knew it, but he was paying the penalty for violating a popular concept—that of the pugilist . . . we should have been cheering [Tunney] to the echo for the perfection of his profession . . ."*

As an adult, Tunney again encountered Shakespeare when he was on route to Europe with the 11th Marine Regiment. One morning a seasick clerk grabbed "his dress shirt to use as a towel," Jay R. Tunney writes, "and when Gene needed that same shirt clean for inspection and couldn't produce it, he was punished with 24 hours of kitchen duty." An avid reader, the clerk apologized and gave Gene Tunney a copy of Shakespeare's *The Winter's Tale*. (Note to Will Rogers: This is the Shakespeare play Tunney read ten times.)

Tunney went on "to read all of Shakespeare's plays, memorizing many passages. [His] introduction to Shakespeare was, at heart, the story of how he mastered much of literature, reading his way through his self-devised college courses."

*In his essay, "My Fights with Jack Dempsey," Tunney writes that as a prizefighter reading had a "practical side." Games of pinochle and listening to jazz were "devices" boxers used to "[get] one's mind off the fight." But he didn't like jazz and, he says, "the mysteries of pinochle were too deep for me. So I resorted to reading as a way to ease the dangerous mental strain during training. I found that books were something in which I could lose myself and get my mind off the future fight."*

*Yes, Tunney says, his critics bothered him: "I was an earnest young man with a proper amount of professional pride." He was also a man of extraordinary self-confidence, strong will, and purpose. "What saved me," he writes, "was my stubborn belief in the correctness of my logic."*

*Tunney also penned an astute article titled "The Ring and the Book." To prove his thesis that "[b]oxing has found recognition in literature since the earliest recorded times," he quotes passages from various literary classics. In Virgil's* Aeneid, *he writes about the "contest between Dares and Entellus and compares their fighting style to two prizefighters of the nineteen- thirties and -forties, King Levinsky and Max Baer. In Homer's* Iliad *he discusses the fight between Nestor and Epeios. The latter's cockiness reminds him of "the barrel-chested bravado of a Jack Sharkey!"*

*The battle in Henry Fielding's* Tom Jones *between Tom and Squire Blifil and Parson Thwackum (great names) turns into "what is known in Negro Harlem as a battle royal [wherein] several pugilists are placed in a ring at the same time and the second the gong sounds each takes a swing at* the man nearest him. Until I became familiar with Fielding, I considered the battle royal a modern development."

Tunney includes Fielding's remarks about a punch to the stomach that reminds him of the March 17, 1897, fight between Bob Fitzsimmons and James J. Corbett. Fitzsimmons knocked out Corbett "with a blow that landed to the pit of the stomach that became known as 'the Fitzsimmons solar plexus punch.'" Fight scenes in William Thackeray's Vanity Fair, William Hazlitt's "The Fight," Victor Hugo's The Man Who Laughs, and Arthur Conan Doyle's Rodney Stone are also discussed.

George Bernard Shaw, Tunney continues, "knows more about boxing and what makes it tick than any of the other litterateurs who have tackled the subject in their books." His 1880 novel, Cashel Byron's Profession, "established [Shaw's] right to be considered an expert in matters pugilistic."

He calls the reader's attention to a passage in which "Mr. Shaw has written a perfect picture of Cashel Byron 'slipping' a punch." Near the end of the round, Cashel's opponent "struck viciously at his opponent's ribs; but Cashel stepped back just out of his reach, and then returned with extraordinary swiftness and dealt him blows from which, with the ropes behind him, he had no room to retreat . . . His attempts to reach [Cashel's] face were greatly to the disadvantage of his own; for Cashel's blows were never so tremendous as when he turned his head deftly out of harm's way, and met his advancing foe with a counter hit."

For Tunney, Shaw proved to be prescient: ". . .the all-knowing Irishman described before I was born, the principal reason for my own effectiveness in the ring—'slipping' a punch and countering, and moving in and out."

## The Meeting

Forty-one years Tunney's senior, Shaw met the boxer in person for the first time in London, in 1928, a few days after Christmas. "They instantly liked and trusted one another, and although competitive with others," writes Jay R. Tunney, "the difference in their ages and their fields of expertise freed them from competing with each other." Shaw "encouraged Tunney's aspirations toward bettering himself intellectually and culturally, and Tunney's presence gave Shaw a brace of reality by carrying him away from what many saw as the fantasy of the theater and the characters that inhabited it . . . Shaw lived in his mind and was attracted by his physical opposite in Gene . . . In Gene, Shaw found a character too incredible to have written about, and this amazed and satisfied him [while] Gene thoroughly enjoyed G.B.S as a person, finding him not only a walking library but a good listener and a thoughtful observer with whom he could share confidences."

Shaw proved to be a true friend. In 1929, when Tunney and his wife Polly were honeymooning in Brioni, an island in the Adriatic, she developed appendicitis and was close to dying. Shaw and his wife Charlotte traveled to the island to comfort the boxer during what was the most difficult time of his life. The Tunneys never forgot the Shaws' kindness, generosity and concern.

## *In Retreat*

While he was training for the second Dempsey fight, Tunney did roadwork daily, usually five miles. Many of those miles were run backwards. He reasoned that if he were to be knocked down, he wanted to be prepared to stay away from Dempsey's punches by backpedaling.

It's 1927. Soldier Field, Chicago. In round seven Tunney *was* floored. Dempsey didn't immediately go to a neutral corner. Referee Dave Barry began counting only *after* Dempsey had gone to a neutral corner, giving Tunney several extra seconds to recover.

Back on his feet at nine, Tunney began retreating. It was "my legs against his," he writes in "My Fights with Jack Dempsey." "The strategy was okay—keep away from him until I was certain that all the effects of the knockdown had worn off."

*John Jarrett quotes sports journalist Trevor Wignal as writing: "Dempsey had frittered away four seconds by refusing to obey the commands of referee Dave Barry. It is the disposition now to blame Barry for the fourteen second count. There is no justification for the charges made against him. He knew, and Dempsey and Tunney knew, that in the event of a knockdown, the man on his feet was obliged to return to a neutral corner."*

*Supreme irony: Dempsey's business managers, Leo P. Flynn and Bill Duffy, met with the Illinois Boxing Commission a few weeks before the rematch and "[insisted] on the strict interpretation and enforcement of the knockdown rule."*

"Why would they do that?" Jarrett asks. After all, Dempsey had a long history of standing over opponents he had knocked down "and belting them again as soon as any part of their body cleared the canvas." Fighting that way "was Dempsey's style, the way he fought from the hobo camps to the championship of the world."

### *In Pursuit*

Jay R. Tunney writes that Dempsey, enraged and frustrated, his "jaw tucked down behind his shoulder and body bent forward," charged after Tunney immediately after the champion was back on his feet. He threw a left hook that Tunney avoided. Dempsey continued chasing Tunney, who "took his time, skating sideways around the ring like a dancer, staying just out of reach."

Then, after Tunney fired a right cross that landed on Dempsey's temple, stunning him, he "changed direction and was now circling to his right." He struck again with another right that landed on Dempsey's chin. Momentarily bewildered, Dempsey stopped and gestured to his foe to stop running, to stand still and fight. That was the moment when Tunney "sensed the truth in Dempsey's eyes: he's discouraged. He's going to lose, and he knows it."

Near the end of the round Tunney spotted an opening and landed what was probably the hardest punch of the fight, "a straight, six inch right that hit directly under Dempsey's heart." (After the fight Dempsey told reporters the punch was so hard that he could barely breathe and thought he'd die.)

For the next three rounds, Tunney out-boxed—and at times even outslugged—the tiring Dempsey. When the decision was announced, Tunney had retained his title by unanimous decision.

In 1928 Tunney defended his title against New Zealander Tom Heeney at Yankee Stadium, in New York. After stopping the challenger in the eleventh round, Gene Tunney announced his retirement from boxing.

# Revisiting Kaletsky, Silver, and Vitale

*Involved*

It was 1963 and Richard Kaletsky, a feisty fifteen-year-old, had just seen Muhammad Ali, then Cassius Clay, win a ten round decision over Doug Jones. He saw the battle, he writes in his memoir, *Ali and Me: Through the Ropes*, "on a closed broadcast, in a smoke-filled hall known as the New Haven Arena."

(And for the record, *Ali and Me* is one of the best books ever written about Ali.)

Kaletsky was fascinated by the young fighter's fistic skills and intrigued by his personality. Then there was his poetry—which "blossomed and flourished in the newspapers. The sports sections were full of the silly, arrogant, but ear-catching rhymes. I couldn't stand it any longer. I had to be involved. It was time to find out what this guy was all about."

The next day he made a collect call to Clay's home in Louisville, Kentucky. "We had a nice chat," says Kaletsky.

*"Nearly twenty years later,"* Kaletsky writes, *"Ali called me at about 1:30 a.m. to thank me for sending him my book. Between calls I had seen him fight 'live, in person' eight times, including the first two bouts with Joe Frazier."*

*Kaletsky has met Ali on several occasions. One of those meetings was "highlighted by a visit to his dressing room immediately following the rematch victory over Leon Spinks, in New Orleans." Also, he has "met many members of Ali's 'corner,' entourage, and family, and attended his 'invitation only' 70$^{th}$ birthday party in Louisville."*

One of the chapters in *Ali and Me* is titled "Bolie Jackson." Bolie who? Well, think back to Rod Serling's *The Twilight Zone* and the

"The Big Tall Wish," an episode aired in 1960. Bolie is an over-the-hill prizefighter whose most loyal fan is ten-year-old Henry, a neighbor's son.

The episode's theme is manifested when the youngster, who still believes in magic, tells Bolie that if he's going to win his next fight, he's must *believe*. Bolie's response is that he can't believe because he's too old and magic, he insists, doesn't exist.

And Bolie, in one of the show's endings, loses: Knocked out because he didn't believe!

So what did Kaletsky learn from "The Big Tall Wish"? Fast forward fourteen years to central Africa, to Zaire, where Ali fought menacing heavyweight champion George Foreman, who was destroying every opponent who stepped into the squared circle against him.

But Kaletsky, you see, believed. He believed Ali would defeat Foreman. He believed in Ali *totally*.

As we know, Ali won by an eighth round knockout and was again world heavyweight champion. And did he win because Kaletsky "beleeeved"? Was it because Kaletsky knew that "Bolie Jackson's little friend was right as could be"?

Here's Kaletsky: "It was good to know that I could make Ali win, for I had learned so well the lesson of Bolie Jackson. Ali's abilities and execution would undeniably help . . . Yet, it was with diamond clarity that I would be the one to assure victory, because 'I beleeeved.' There I was—a twenty-six year old, slightly overfed, well-educated, upper middle-class, suitably employed male—repeating over and over and over, 'I believe, I believe.' No put-on."

### *Too Tough, Too Skilled*

Noted boxing historian Mike Silver's *The Arc of Boxing: The Rise and Decline of the Sweet* is both a tribute to boxing's Golden Age (1920 to 1950), and a compelling criticism of the sport as it is today.

And when Silver says that anyone who "who reads [*The Arc of Boxing*] will never look at a boxing match the same way again," he's correct.

Believe me, there's something informative and valuable on every page.

Two well-known middleweight champions of the recent past were Carlos Monzon and Marvin Hagler. Silver says that Monzon (87-3-9 KO 59), middleweight titleholder from 1970 to 1977, defeated only two worthy challengers, Benny Briscoe and Emile Griffith, who was past his prime. His "toughest opponent was formidable Rodrigo Valdez." His other defenses "involved ordinary club fighters such as Jean Claude Bouttier, Tom Bogs, Tony Mundine, Fraser Scott, and Tony Licata."

Hagler (63-3-2 KO 52) dominated the middleweight division from 1980 to 1987 and retired after a disputed twelve round decision loss to Sugar Ray Leonard. During Hagler's reign, Silver says, "Vito Antufermo, Alan Minter, and Tommy Hearns were his most worthy contenders."

(I'd include John "The Beast" Mugabi on that list. Slightly behind before he was stopped in the eleventh round of their 1986 title bout, the hard punching Ugandan gave the Brockton destroyer a competitive fight. It was Mugabi, boxing historian and writer George Kimball says, who first "unmasked the very erosion of skills that Leonard would exploit later.")

Supporting Silver's thesis that fighters of the Golden Age were superior to today's fighters is boxing manager Mike Capriano Jr: "Marvin Hagler would have been too good for the middleweights of today. Yet people we knew in the 1940s and early 50s would outbox Marvin Hagler. From our experience, the people we saw looked better, were better, and had an understanding of what to do. And they were better inside fighters."

Before they won their middleweight titles, Monzon had eighty-one fights; Hagler, fifty-four. Silver contends that if they were fighting during the Golden Age, they "would have been ranking contenders and perhaps even won the title, but they would never have been as dominant as they were in their time. Until the early 1960s the middleweight division was just too damn tough."

Bernard Hopkins won the middleweight title at age thirty, and before Jermaine Taylor defeated him in 2005, he defended it "a record breaking" twenty times. A notable achievement. But Silver contends the middleweight champions of the Golden Age—Sugar Ray Robinson, Mickey Walker, Harry Greb, and Jake LaMotta—"could never have defended their titles 20 times over 10 years against the kind of brutal competition that populated the middleweight division from the 1920s to the 1950s." Doubtless Hopkins is talented. But as middleweight champion he was "surrounded by a sea of mediocrity" and "benefitted from the worst assortment of challengers ever faced by a middleweight or light heavyweight champion since the advent of boxing gloves."

### *Name Change*

Fighting the feared Stanley Ketchel for the undisputed middleweight championship in 1908, Hugo Kelly, one of the lesser-known fighters about whom Rolando Vitale writes in *The Real Rockys:" A History of the Golden Age of Italian Americans in Boxing 1900-1950,* was holding his own, until he became overconfident and was stopped in the third round.

Shortly after Ketchel was murdered in 1910, an elimination series to determine his successor was announced. Kelly's decision over Frank Klaus and several other notable opponents enabled him, Vitale writes, "to claim the world title again." In that same year, after several more impressive wins, he "lost his title claim to Jack Dillon when he was dispatched in three rounds . . . That was the end of Hugo Kelly as a main attraction."

Born in the village of Vitiana, a province of Lucca, Italy, Kelly's birth name was Ugo Geno Micheli. He was world middleweight claimant from 1905 to 1907 and again in 1910. Exactly how good was he? The *Chicago Tribune*'s 1909 edition described him as "'the cleverest middleweight in the world, feline in movements, knows every turn in the road.'"

There were several reasons why Kelly and "so many Italian American fighters [changed] their names . . . during the period from 1880 to 1910," Vitale writes. One reason was the overt "discrimination, prejudice, and hostility" they encountered regularly. Another reason: The often deliberately "irritating and disrespectful practice of mispronounced and misspelt family names." Financially, a name change offered "a better chance of regular work and higher billing on a boxing card."

That many Italian fighters anglicized their surnames didn't mean "they lost sight of who they were," Vitale continues. It was, rather, a way "to integrate into American society and embrace the host culture [but] did not mean that they traded in their Italianate beliefs and values."

Of course bigotry was not "solely experienced by the Italian cohort, and "the practice of assumed names was not exclusively reserved for the Italian or restricted to boxing. The Italian followed an earlier trend with both the Jewish and German cohorts embracing Irish and anglicized tags."

*Mushy Callahan wasn't always Mushy Callahan. Born Vincent Morris Scheer in New York City, in 1905, he was world light welterweight champion from 1926 until 1930, and fought mostly on the West Coast. As for his name change, with an artful wink Jack Cavanaugh, in* Tunney, *writes, "He took [it] to an extreme when he converted to Catholicism, married an Irish girl in 1934, and had a son who became a priest."*

According to boxing historian and film buff Mike Silver (The Arc of Boxing), *"Mushy Callahan's first name is a derivation of his Hebrew name 'Moishe,' which transmogrified into Mushy, his [boxing name].*

*"By the way, Mushy is only one of three Jewish pro boxers who later converted to another religion. But we made up for the loss by having the original Golden Boy, Art Aragon, convert to Judaism after he married a Jewish girl. I got this straight from the Golden Boy himself."*

*A good counterpuncher and excellent defensive fighter, Callahan fought from 1923 to 1932 and retired with a record of 48-16-3 (KO 21). In retirement, he lived Los Angeles and became a boxing judge and referee. He was a boxing adviser and choreographer on a number of boxing themed films as well. He also acted in movies, usually unbilled. Some of them were* They Made Me a Criminal *(1939),* Ironman *(1951),* The War of the Worlds *(1953),* The Bamboo Prison *(1954), and* Birdman of Alcatraz *(1962).*

*His most notable film was* House of Strangers *(1949), a forgotten gem from Twentieth Century Fox. Told in flashback and directed by Joseph Mankiewicz, the film starred Edward G. Robinson, Richard Conte, Susan Hayward, and Luther Adler. Callahan plays a referee in this deftly crafted film about a ruthless banker who lets his loyal son take a prison rap for him.*

Vitale became a fan of boxing after he and his father watched Muhammad Ali fight Joe Frazier on television. Since that time, he says, "I have been spellbound by the compelling unpredictability of outcome and by the boxer's level of sacrifice and professionalism, seldom matched by other competing disciplines . . . I felt an enormous responsibility to bring together the forgotten age of the Italian American boxing experience."

With a real genius for Italian and American history, and Italian American boxers, as proved in nine reader-friendly chapters and twenty-one comprehensive appendices, Vitale's *The Real Rockys* is erudite, readable, and painstakingly researched.

**Lengthier reviews of *Ali and Me* (Issue 115/ September 2013*)*, *The Arc of Boxing* (Issue 118/ June 2013), and *The Real Rockys* (Issue 124. December 2014) were first published in the *International Boxing Research Organization Journal*.**

# Where Have You Gone, Court Sheppard?

As we already know, sports journalist Paul Gallico sparred with heavyweight champion Jack Dempsey in 1923. Young and fearless, he wanted to experience first-hand how hard the heavyweight champion could punch and then write about for his newspaper, the New York *Daily News*.

In less than two minutes he hit the deck twice, emerging from the experience with a terrible headache and numerous bruises. (Surprise!)

Comfortably seated in his office while Gallico was being pummeled (another surprise), the reporter's editor congratulated him the next day: "Nice work, kid. Hope you didn't get hurt. Even if you did, you earned yourself a by-line."

While filming 1942's *Gentleman Jim*, actor Errol Flynn sparred with heavyweight Jack Roper, who was hired to train the actor for his role as heavyweight champion James J. Corbett. If you believe what Flynn says in his autobiography, *My Wicked, Wicked Ways*, Roper accidentally knocked him out not once, not twice, but three times in one day. (Yes, Flynn sparred with Roper and was flattened at least once, maybe even twice. But three times? Do I sense a tad of bullshit on Flynn's part?)

Jack Roper's claim to fame was that he fought heavyweight champ Joe Louis in a title bout in 1939 at Wrigley Field, in Los Angeles. Though game, Roper didn't make it out of the first round. When reporters later asked him what happened, he uttered the classic line, "I zigged when I should have zagged."

(While Roper was actively boxing, he appeared unbilled with Myrna Loy and William Powell in one of their *Thin Man* films. When he retired he acted in several *Joe Palooka* motion pictures.)

At Stillman's Gym in New York, in 1959, participatory journalist George Plimpton laced up "the leather mittens"—sports journalist Gene Ward's term—against light heavyweight champion Archie Moore. He did so with the intention of making the experience the centerpiece of *Shadow Box*, the book he was writing.

Plimpton went into serious training for the Moore exhibition. He hired experienced George Brown as his trainer. He sparred. Worked the heavy bag. Jumped rope. Did road work usually through Central Park. He admits that on some runs he'd daydream, and that's when Brown would remind him "to work up a controlled rage against Archie Moore, seeing him always in my mind's eye, shadow boxing as if his presence were just beyond reach . . ."

Scheduled for three rounds, as "the event" progressed the savvy Brown noticed that Moore's "moods seemed to change . . . that the fighter was getting testy . . ." Brown knew he'd have to do something before Plimpton was hurt. And he did! Much later Plimpton learned "that [in the final round] Brown had reached down and advanced the hand of the time clock. The bell clanged sharply with a good minute to go."

When the sparring session was over, Plimpton had a bloody (maybe broken) nose, courtesy of a Moore left jab in round one. But like Gallico before him, he had a story.

In his breakout role, Oscar winner Paul Newman portrayed Rocky Graziano in the 1956 film adaptation of the former middleweight champion's autobiography, *Somebody Up There Likes Me*.

During the 1940s Tony Zale, who battled Graziano in three title bouts, all wars, had been hired to play himself in the film. What happened to Newman is recounted by Thad Zale, Tony Zale's nephew, and Clay Moyle in their impressive biography *Tony Zale: Man of Steel*.

The camera rolled. Newman and Tony advanced toward each other. (Both men had been instructed to pull their punches.)

It wasn't long before Zale fired a body blow, and Newman, according to Thad Zale and Moyle, "suddenly grabbed his side, and doubled over, and yelled, 'Whoa man! You trying to kill me?'" The former great middleweight champion stopped punching, apologized, and the action resumed. After a minute Zale "shook up Paul with a couple of light hooks to the head and then came in with a right to the stomach that sent the latter to the canvas gasping for air." The director, Robert Wise, immediately called cut.

That Newman became too involved in the role isn't an exaggeration. He was a method actor. Method actors *become* their characters. So for the film Newman *became* the hard hitting Graziano. "During the first two practice rounds, Tony had repeatedly warned Paul not to hit him full steam. Unfortunately, for both parties, the young actor was either unable, or chose not to follow that advice," Thad Zale and Moyle write. "Years later Tony would say that he thought Newman caused the problem by trying to deliver too many hard blows."

The incident ended Zale's thespian aspirations, but he was paid for the time he put in, and of course the film's producers immediately went looking for a replacement: "[They] found a less dangerous, professional actor named Court Sheppard for [Newman] to work against. Court played the part perfectly and they used a lot of good old fashioned makeup to help him resemble Tony."

For Sheppard, born Courtland Schultz in 1914, the Zale role wasn't his first acting assignment. He began performing in 1949 and made his last film, *The Jayhawkers*, in 1959. In addition to his role in *Somebody Up There Likes Me*, he appeared uncredited in a number of other boxing movies: *The Leather Saint*, *Champion*, *Right Cross*, *The Fight*, and *Tennessee Champ*.

Sheppard boxed professionally as a light heavyweight from 1937 until 1941 and compiled a 14-2-3 record, with ten knockouts. Even more impressive is that, according to *boxrec.com*, he was the St. Louis Golden Gloves Champion in 1936, defeating—brace yourself—Archie Moore in the finals.

**"Where Have You Gone, Court Sheppard?" was originally published in *the International Boxing Research Organization Journal (Issue 126/ June 2015)*.**

# Sonny's Visit

Journalist Joe Flaherty asks, "Was Sonny [Liston] Satan?" His answer: "Not really, but he'd make a helluva understudy."

Sonny Liston wasn't the bastard many people said he was. Writer and social activist Claude Brown speaks not only for himself but also for many others when he says Liston "was the only man alive who could have quelled the Watts riots."

Flaherty spent five days with Liston in Los Angeles, where the former heavyweight champion was beginning his comeback in 1966. Liston died four years later and Flaherty wrote "Amen to Sonny" in 1974 for *The Village Voice.* He says that Liston, "the son of an Arkansas sharecropper who fathered twenty-five children" and beat him every day, terrified much of white America because he epitomized "the menacing black man who invaded the subway of our souls at four in the morning." After all, he had been arrested at least twenty times. More, his resume included honing his thuggery skills for the union, cracking heads for the mob, attacking police officers, and driving drunk.

*When Michael Feldman was eight years old and the only Jewish child at Poland Community School in Poland, Maine, he learned that Sonny Liston was in town training for his May 25, 1965, rematch with heavyweight champion Muhammad Ali, at St. Dominic's Hall in Lewiston, Maine. Liston was staying at the Poland Spring House, owned by Feldman's grandfather Saul.*

*With the enviable courage, directness, and innocence possessed by youngsters, Feldman approached Liston and asked him if he'd visit his school. And the former champion agreed,* writes Mark Emmert for the Maine Sunday Telegram, *"thus creating probably the most unlikely scene ever witnessed at [the school]."*

*On the morning of Liston's arrival young Feldman led "a large and impeccably dressed black man and his entourage from one classroom to another." Feldman recalled that Liston's reputation was that of a hardened ex-con, gruff and violent, but with the school children and staff he was relaxed, "easygoing and friendly. . ." Feldman's family enjoyed his company, too, and set aside a private dining room for him, "and invariably [he] invited the Feldmans to join him, which they did."*

Flaherty says that Liston is "only a minor-leaguer in evil compared to the sport at which he toiled, a crude crusher in the domain of charlatans. He wasn't allowed a license in New York State, and indeed, he wasn't even allowed to be introduced in the Garden before fights . . . Of course, this is the same New York that denied Ali a license for being unpatriotic until a $10,000,000 gate appeared on the horizon and transformed him from a traitor to a pugilistic Patton. Ah, what the green can do for the old red, white, and blue."

Flaherty concludes: "I'm not pleading for his life-style—a bastard, maybe, or, perhaps more fair, he did bastardly deeds. But he should be judged in context. He was better than the sport he practiced and the men who rule it."

*Liston lost the fight against Ali. Witnessed by about twenty five hundred spectators, it was over in one round.* "The timer has said that the fight lasted one minute," says Nat Fleischer, founder and editor of Ring magazine. "The people who took the movies for the TV say it was 1:48. I say it was 1:42—1:32 plus ten seconds, by my watch." The punch landed on Liston's left jaw and was a "short right hand blow" and "far from devastating . . . and the jeers that followed testified to the attitude of those who felt they had been bilked."

*According to Emmert, Feldman recalls that he was too upset to attend school the day after the fight:* "I was a little emotional about him losing the fight, as everybody here was."

*Over the years Feldman, now a real estate broker, tried unsuccessfully to contact Liston:* "I just think it was the turn that his life took. And I'm sure he was embarrassed about the situation . . . To have that reputation as a tough guy and then roll in here and charm people like that. I considered him just a friend, a wonderful guest and friend. We all did."

*Liston died on December 30, 1970. The cause was "drug-related heart failure," Emmert writes, but there were rumors that he was murdered.*

*He's buried in Paradise Memorial Gardens in Las Vegas.*

Prior to the second Ali fight, Liston's chin was never questioned. In *The Gods of War,* Springs Toledo writes that Mike DeJohn and Cleveland Williams, two thunderous punchers, tagged Liston on the jaw several times with little effect. In 1959 Liston stopped Williams in the third round. He halted DeJohn, also in 1959, in the sixth round. In Liston's rematch against Williams a year later, he ended matters in the second round.

There was also Marty Marshall, the first fighter to defeat Liston. He broke Liston's jaw with a right cross early in their 1954 ten rounder. Toledo quotes Marshall as saying, "I never knew he was hurt. You hit him with your Sunday punch but he don't grunt, flinch, or blink. . ."

In boxing's most famous photograph, Ali is standing over the fallen Liston on that fabled 1965 night in Lewistown, Maine, yelling at him to get up and fight. He's angry that Liston had (perhaps) taken a dive. "It was only later that Ali and company came up with the 'anchor punch' spin for posterity's sake," Toledo says. "It's understandable. A dive taints both fighters."

After the Ali debacle, Liston won fourteen consecutive fights. A resurgence? Maybe. But on December 6, 1969, along came reality in the person of Leotis Martin of Philadelphia. Liston pounded him for eight rounds, but in the ninth Martin caught him with a right, a left, and another right to the head. Liston went down. Announcing the nationally televised fight was Howard Cosell, who almost had a green hemorrhage as Liston was counted out.

Liston was back in the ring on June 29, 1970, against Chuck Wepner of Bayonne, New Jersey. Liston battered him savagely and the fight was stopped after nine rounds. "Sonny had hopes that this, his fiftieth victory, would qualify him for more lucrative bouts. It was not to be," writes Toledo. Six months later "the Grim Reaper showed up instead, tapping him on one of those massive shoulders."

# Did Harry Save Rocky?

1

It was the fight of Archie Moore's life. He was ecstatic because he was finally getting a shot at Rocky Marciano's crown. After all, as Mike Fitzgerald says in *The Ageless Warrior: The Life of Boxing Legend Archie Moore*, "The heavyweight championship is boxing's grandest title . . ."

To get the fight Archie's brain trust had to wage "an unprecedented public relations blitz . . . 127 letters were sent to writers and sports journalists around the county. Eventually the mailing list grew to 500 newspapers and magazines. Archie himself wrote as many as 150 letters daily." He made television appearances. Perhaps best of all "he even popped up unannounced to pester Marciano when Rocky was [a guest referee] at Turner's Arena in Washington, D.C . . ."

And the blitz worked! He and Marciano signed to fight for Rocky's title on September 21, 1955, at New York's Yankee Stadium.

Sixty thousand fans showed up, and the gate was bigger than Marciano's two previous title defenses against Jersey Joe Walcott and Ezzard Charles.

In the ninth round Moore finally caved in under Marciano's furious assault. Speaking to the press after the fight, Moore, who knocked Marciano down in the second round, said that had he paced himself and been more patient, he might have been better able to handle the champion in the later rounds.

In his interview with Peter Heller, author of *In This Corner . . .* ! Moore was critical of referee Harry Kessler. Kessler was "callous" and "had no business handling a match of this magnitude. Harry Kessler was the guy who saved Rocky Marciano. Kessler was so excited. He had no business refereeing that match because he was too excitable. He didn't know what to do. He grabbed Marciano's gloves and began to wipe Marciano's gloves and look over his shoulder and gave Marciano a count standing up. If a man is on his feet, he is automatically a target. This is what was told to us at noon, but he forgot all about it and he began to rub his gloves and then gave him a snatch, you know, kind of to pull him to. I'll never forget it. It cost me the heavyweight title. I never take any credit from Rocky, because I think Rocky was one of the greatest fighters that ever lived, and one of the nicest fellows that ever lived."

Moore and Kessler have a history dating back to 1952, when Moore dethroned light heavyweight champion Joey Maxim. Kessler refereed the one-sided affair. When it was over, Judges Fred Connell and Howard Hess scored it overwhelmingly for Moore, 82-68 and 87-63, respectively. The AP had Moore winning, 84-66. (I scored it it 85-65, Moore.) In rounds, one judge gave Moore thirteen, Maxim one, and one even. The other judge saw it nine for Moore, two for Maxim, and four even.

Harry Kessler watched a different fight. His scorecard had Moore winning, 76-74. In rounds he scored the fight a draw, seven for each fighter and one even.

In his book Mike Fitzgerald quotes Kessler, who, in 1982's *The Millionaire Referee,* his autobiography, writes that perhaps the judges "subconsciously had tried to reimburse Archie for the fourth round, which I had taken away [for] the low blow fouls. Not that it mattered. I scored the fight the way I saw it, and so did they."

2

At New York's Madison Square Garden, the year before his fight with Marciano, Moore defended his light heavyweight title against longtime rival Harold Johnson, a superb boxer-puncher and one of the best fighters of the era.

In the tenth round Johnson, who was on a twelve bout winning streak, dropped Moore, writes A.J. Liebling in *The Sweet Science,* "with a beautiful overhand right to the left side of the head . . ."

The knockdown was "so unexpected, so unprecedented that even the referee, Ruby Goldstein, lost his head. Goldstein's first impulse must have been to help Moore to his feet and apologize on behalf of the management, but he checked it in time and began to count."

Moore was up at three, but Goldstein continued counting to five. "In the lost two seconds, Johnson might have hit Moore a couple more licks if Goldstein hadn't been there and the challenger could have prevailed on himself to take the initiative."

In round fourteen, Moore, knowing he was behind on points, went after Johnson. A bombardment of punches sent Johnson to the canvas. He was on his feet at the count of four, and though "no mandatory eight count was called for by the rules," Fitzgerald writes, "Goldstein administered one anyway."

Johnson had been given a four second respite.

Liebling writes that Archie tore into him, "played a solo on [his] head, and Goldstein stopped the fight."

# Joe and Terrible Tony

***Waking the Lion***

Heavyweight champion Joe Louis would defend his title for the seventh time against "Two Ton" Tony Galento on June 29, 1939, in Yankee Stadium. Fearless and the owner of a powerful left hook, Galento was rarely in condition. He was also one of the dirtiest fighters of the era. A New York *Times*' article several weeks before the fight reported that when "Galento fought Auturo Godoy some time back, all sorts of tactics not endorsed by the books—gouging, elbowing, and butting—were brought into use." Referee Arthur Donovan "took the attitude that if the boys didn't mind, neither did he." (Guess who would referee Louis' fight against Galento?)

Louis knew of Galento's disregard for the rules and made it clear if the challenger fought dirty, he'd respond in kind.

Before almost 35,000 fans, Galento, whom New York *Times*' James P. Dawson described as "the despised, roly-poly, unorthodox, uncontrollable heavyweight," staggered Louis is the opening round with a left hook to the jaw.

Louis roared back in the second session, flooring Galento—who outweighed the champion by thirty pounds, though he was four or five inches shorter—for a two count.

In the third round, the challenger again landed a left hook to the champion's jaw. And suddenly the crowd was on its feet! The contender, whom no one had given a chance to win, had put the great Joe Louis on the canvas! The punch "landed in grazing fashion," Dawson writes. "Its full force was spent before it connected because of Tony's shorter reach and the fact that Louis was going away from the punch." More embarrassed than hurt, Louis was up at the count of two.

Louis knew he was in a fight and better end it soon! So the champion went on the hunt and Dominick Anthony Galento was his prey. Not good news for the challenger.

It was over in the fourth round: Louis attacked, battering the badly bleeding Galento mercilessly. "He was knocked from pillar to post, bashed in the face, stabbed in the ribs and draped on the ropes, helpless but game," John Kieran of the *Times* would write. That's when referee Donovan "mercifully stopped the incipient assassination . . . But there had been some thrilling moments in that fight and the gallant and globular Galento did himself proud."

(By popular request I'll throw in a quote from Shakespeare here. It fits both Galento and Louis. It's from *Twelfth Night*: "If one should be a prey, how much better/ To fall before the lion than the wolf!")

*In* Joe Louis: Black Champion in White America, *Chris Mead writes, "Only a brave man would have said the things Galento said about Louis before the fight." Brave man? Come on, Chris. Galento's remarks crossed the line. They were personal and infuriated the champion, who never had a bad word to say about an opponent.*

*Afterward Louis admitted he was angrier at Galento than at any other opponent; that he wanted to severely punish the saloon owner and former bouncer for ten rounds, then knock him out. That Louis changed his mind after being dropped and went for the knockout was a good thing for Galento. Probably saved him from permanent injury.*

After the fight, Dawson writes, "Everybody was singing the praises of the barrel-chested heavyweight from New Jersey . . . And the leader of the chorus was Louis." The champion said Galento was one of the hardest punchers he ever faced; that he was a good fighter and stood up under some tremendous punches.

*Mr. Galento, you didn't learn anything from the beating Joe Louis gave you, did you? I say that because several days before your 1940 fight with Max Baer, according to John Jarrett in* Dynamite Gloves, *you "infuriated [him] by mailing him a postcard showing two scraggly unkempt bears pawing one another, across which [you] had written, 'Max and Buddy, and this goes for your whole family, you bums'" And at the weigh-in you called Baer yellow, just for good measure.*

*Once the fight got underway, you butted and bit and heeled and "ripped at Max's face with the laces of [your] gloves. But Baer fought back with a savage fury not seen since the [Max] Schmeling fight and chopped [you] to pieces." When he finished with you, you were "a bleeding, blubbering hulk, sitting in his corner" unable to answer the bell for round eight.*

### Curious Article

In Jack Kofoed's bizarre article about Galento, "Tony the Terror," he says Galento is "a swell guy" outside the ring and is "the John L. Sullivan of the modern world and John L. Sullivan was a figure without parallel in all the gaudy history of the ring."

I have no doubt that Galento was a good guy when he wasn't fighting. (Most fighters are.) As for the Sullivan comparison, Kofoed admits Galento isn't as good as Sullivan, "no matter what he himself may say on the subject. But Galento has the same braggadocio and dash of the old-timer." And true, there wasn't "an ounce of quit in [Galento's] 230 pounds of fighting flesh."

But Galento was "an in and outer. Nobody knew what he would do. He didn't know himself. He always tried, but one night he would have it, and another he wouldn't. That had been his fault from the beginning . . ." More, "the guy has pulled more screwball stuff than anybody who can be listed offhand." (Supposedly on a bet he once ate fifty hot dogs before a fight.)

Unsettling about Kofoed's piece is what he says about Galento's 1932 fight with Ernie Schaaf, at the time the world's third ranking heavyweight. When Schaaf fought Galento the "battle was one of those Pier A brawls. Anything went." Schaaf won the decision, but a month later lost to Stanley Poreda. Then two months after that Max Baer defeated him. (There were two seconds remaining in the tenth round when Baer knocked him unconscious. Saved by the bell, Schaaf was dragged unconscious to his corner.) Later in the year he fought Primo Carnera. He was stopped in the thirteenth round and never regained consciousness. Kofoed writes that "it might be noted, this top notch fighter [Schaaf] began to slip very badly after he came out of that battle [with Galento]."

### Rashomon*

In Louis' title defense against challenger Billy Conn on June 18, 1941, at New York's Polo Grounds, the champion knew he was behind going into round twelve.

Conn's trainer, Johnny Ray, believed his fighter "would surely win the decision and take the title that way." But Conn was positive "he could knock Louis out, and a knockout would bring greater glory." Well, Conn was the one knocked out. According to Mead, "Ethnic-conscious sportswriters found an easy explanation for Conn's defeat. They wrote that Conn's Irish blood got him caught up in the fighting and made him go for the knockout instead of boxing cautiously and going for the decision."

And Conn's trainer agreed, stating that if Conn was Jewish, not Irish, he'd be world heavyweight champion.

Writing for the *International Boxing Research Organization Journal*, Bobby Franklin saw it this way: Moments before the bell ended round twelve, "Louis hit Conn with a very hard right hand. Billy's knees buckled, and after the bell sounded he seemed disorientated and was confused . . . This was the beginning of the end." (It was!)

Several years later, writes Randy Roberts in *Joe Louis*, Louis and Conn had become friends. Always ready with a wisecrack, Conn asked Joe why he didn't let him borrow the title for a little while. Maybe six months. Equally quick witted, Louis replied, Billy, you had it for twelve rounds. Why do you think you could've kept it longer?

*Billy Conn and his father-in-law, "Greenfield" Jimmy Smith, didn't see eye-to-eye about anything. It was 1942, early evening and, says Roberts, they attempted "to patch up their difference, but it did not take." The Smith's kitchen was the scene, and what began as a discussion quickly escalated into a shouting match. When "Greenfield" Jimmy said he could still beat the crap out of his son-in-law, Conn exploded. "...at the time he was sitting on the stove. He came off it like he had been burned and threw a looping left hook." It landed on his father-in-law's skull but "broke Conn's hand," costing him "his chance for a big-money rematch with Louis before they both went off to war."*

According to Roberts, whenever Louis met with Conn, he'd ask if "Greenfield" Jimmy was still beating the crap out of him.

**\*Rashomon (1950) is the Japanese film in which several people have different interpretations of the same event.**

# Long Overdue

A masterful writer with a firm understanding of boxing and Gene Tunney's life, Jack Cavanaugh's book, *Tunney: Boxing's Brainiest Champ and His Upset of the Great Jack Dempsey*, is about Tunney and many of the individuals who played a large part in his life: Jack Dempsey, with whom he'll be forever linked; Polly Lauder Tunney, his beautiful socialite wife; Harry Greb, whom he fought five times; and Tex Rickard, one of boxing's greatest promoters.

Cavanaugh succeeds admirably in building a solid case for Tunney's place among professional boxing's elite.

### *Always Several Steps Ahead*

Tunney regarded Tommy Gibbons "as the best heavyweight in the world after Dempsey" and wanted him as his next opponent. He knew he'd be assured of a title fight with Dempsey if he defeated the talented Gibbons, whom Dempsey had outpointed in their 1923 title fight in Shelby, Montana.

Gibbons, who had never been knocked down or out, thwarted Dempsey's "aggression," Cavanaugh writes, "by clinching repeatedly and . . . eluding most of Dempsey's punches." Bigger and stronger than Gibbons, Dempsey "took advantage of the numerous clinches to inflict rabbit punches . . . while also roughing up Gibbons inside."

Eddie Kane, Gibbons' manager, knew of Dempsey's "penchant for such tactics [and] had asked that they be prohibited, as they were in most states." But Jack Kearns, Dempsey's manager, had more influence than Kane and "had demanded as a precondition to any contract that both rabbit punches and kidney punches be permitted." (Also, Jim Daugherty, Kearns' friend and favorite referee, was assigned as referee.

So there was absolutely no chance "that Dempsey was going to be disqualified for a low blow, short of a punch to the groin that might incapacitate Gibbons.")

Always several steps ahead of his opponents, Tunney had been studying Gibbons' style for almost two years. He "knew that the veteran from Saint Paul's most effective tactic was to feint twice with his right hand . . . and, then, with his opponent protecting his head, Gibbons would move his left foot forward, drop his head and whip across a powerful left hook to the head or to the body."

Tunney and Gibbons fought on June 5, 1925, in New York, and it didn't take long for Gibbons to realize "that Tunney was anticipating not only his lethal left off the double-feint, but almost everything else he planned to do." Forty thousand fans watched Tunney, fighting methodically, completely overwhelm Gibbons and stop the St. Paul veteran in the twelfth. The defeat, according to James P. Dawson of the New York *Times*, "Closed the championship hopes of Gibbons," who wisely announced his retirement after the fight.

### *Their First Meeting*

Cavanaugh describes the first time Tunney met Dempsey, which wasn't in the ring but on a Hudson River ferry in 1920: "Hesitant, but exuding—perhaps feigning—confidence, Tunney strode over to where Dempsey was reading a newspaper . . ." and introduced himself to the champion.

The two men shook hands and Dempsey asked how Tunney was feeling. Tunney said he hurt the middle knuckle finger of his right hand boxing in France over a year ago, and it hasn't healed. Dempsey took "Tunney's right hand in his own huge hands—much bigger than Tunney's—[and] looked at it closely while gently feeling the sore knuckle. Tunney was touched by Dempsey's warmth, and his obvious interest in the injured knuckle."

He advised Tunney, Cavanaugh writes, "to put black tape on the good knuckles on both sides of the injured one before putting on his boxing gloves. That way, Dempsey explained, when Tunney landed a punch with his right hand, the padding of tape on the good knuckles would absorb the shock from the injured knuckle."

Tunney never forgot "Dempsey's friendliness, gentleness, and willingness to help a fellow fighter, even a relatively obscure one."

### The Letter

Someone in Dempsey's camp—Cavanaugh believes it was "most likely Leo Flynn," Dempsey's manager at the time—sent a letter to the *Chicago Herald-Examiner* that was published three days before Tunney's 1927 rematch with Dempsey.

It began with the standard salutation, then went on to say that ringside officials, backed by people in high places, would make certain Tunney won the decision. Too, there were claims that the first bout, a year ago, was fixed because Tunney and his manager, Billy Gibson, were in cahoots with shady Philadelphia bootlegger Max 'Boo Boo' Hoff.

Tunney's response began the way Dempsey's did. Then he stated Dempsey's letter was a plea for sympathy from the public. Tunney closed with the equivalent of a hard left hook to the stomach: He wrote he had written his letter himself.

There was an issue regarding the referee, too: ". . . right up to the opening bell of the rematch," Cavanaugh writes, "suspicions lingered about whether the fight was on the level." The referee would not "be picked until just before the bout started, so that mobsters would have a difficult time getting to the referee in time to bribe him." There was also concern that "maybe someone connected with the Illinois Athletic Commission, which was to appoint the referee and two judges, just might tip off a gambler or a mobster for a hefty price as to whom the referee and judges were to be."

There was more. If the fight went ten rounds to a decision, instead of having the judges and referee "give their decisions to the ring announcer, the commissioner said the cards of the three officials would first have to be handed to the commissioners at ringside." If the commissioners thought the decision wasn't fair, "they would reserve the right to reverse it—which would be unprecedented—or decide not to have a decision rendered at all."

When Dempsey's and Tunney's camps learned of the commission's inexplicable rulings, they were stunned. But soon it dawned on them that, after all, "this was Chicago" in the 1920s—"that toddling town," as Mr. Frank Sinatra sings—"and anything could happen."

**"Long Overdue" was first published in the *International Boxing Research Organization Journal* (Issue 127/ September, 2015).**

# Sources

Allen, Frederick Lewis. *Only Yesterday: An Informal History of the 1920s* (1931)
Basinger, Jeanine. *Silent Stars*. Alfred A. Knopf (1999)
Blake, James Carlos. *The Killings of Stanley Ketchel* (2005)
Bosworth, Patricia. Marlon *Brando* (2001)
Cavanaugh, Jack. *Tunney: Boxing's Brainiest Champ and His Upset of the Great Jack Dempsey* (2006)
Douglas, Kirk. *The Ragman's Son* (1988)
Eisner, Lotte H. "A Witness Speaks": *Portrait of an Anti-Star*, edited by Roland Jaccard (1996)
Fitch, Jerry. *James Louis Bivins: The Man Who Would Be Champion (2011)*
Fitzgerald, Mike. The Ageless Warrior (2004)
Gaiman, Neil. *Trigger Warning* (2015)
Heinz, W.C. "Brockton's Boy": *The Top of His Game*, edited by Bill Littlefield (2015)
Heinz, W.C. "The Same Person Twice": *Once They Heard the Cheers* (1979)
Heller, Peter. *In this Corner . . . !* (1973)
Jarrett, John. *Gene Tunney: The Golden Guy Who Licked Jack Dempsey Twice* (2003)
Jarrett, John. *Dynamite Gloves: The Fighting Lives of Boxing's Big Punchers (2001)*
Kahn, Roger. *A Flame of Pure Fire: Jack Dempsey and the Roaring Twenties* (1999)
Kaletsky, Richard. *Ali and Me: Through the Ropes* (1982)
LaMotta, Jake. *Raging Bull* (1970)
Lopate, Phillip. *Portrait inside My Head* (2013)
Moretti, Sandro. *Mickey Rourke: Wrestling with Demons* (2010)

Moyle, Clay. *Billy Miske: The St. Paul Thunderbolt* (2011)
Munn, Michael. *Kirk Douglas: The Man—The Actor* (1985)
Nagler, Barney. "The Story of a Champion": *The Italian Stallions*, edited by Thomas Hauser and Stephen Brunt (2003)
Oates, Joyce Carol. *On Boxing* (1987)
Parris, Barry. *Louise Brooks: A Biography* (1989)
Povich, Shirley. *All Those Mornings . . . At the Post* (2005)
Roberts, Randy. *Jack Dempsey: The Manassa Mauler* (1979)
Roberts, Randy. *Joe Louis* (2010)
Shakespeare, William. *Twelfth Night* (Signet edition, 1987)
Silver, Mike. *The Arc of Boxing: The Rise and Decline of the Sweet Science* (2008)
Taraborelli, J. Randy. *Sinatra: Behind the Legend* (1997)
Toledo, Springs. *The Gods of War: Boxing Essays* (2013)
Tunney, Gene. "The Long Count": *At the Fights*, edited by George Kimball and John Schulian (2011)
Tunney, Jay R. *The Prizefighter and the Playwright: Gene Tunney and George Bernard Shaw* (2010)
Vitale, Rolando. *The Real Rockys: A History of the Golden Age of Italian Americans in Boxing* 1900-1955 (2015)
Ward, Geoffrey C. *Unforgivable Blackness: The Rise and Fall of Jack Johnson* (2005)

**Magazines and Newspapers**
*Boxing Illustrated* (September 1951)
*Maine Sunday Telegram* (2015)
*The Ring* (August 1951)
*Golden Book Magazine* (April 1934)
*International Boxing Research Organization Journal*
*The Sun Herald* (1982)
*The Lowell News* (May 20, 1949)
*Fight Stories* (Spring 1940)

www.ingramcontent.com/pod-product-compliance
Lightning Source LLC
Chambersburg PA
CBHW030908180526
45163CB00004B/1753